HARVESTING RAINWATER FOR YOUR HOME

DESIGN, INSTALL, AND MAINTAIN A SELF-SUFFICIENT WATER COLLECTION AND STORAGE SYSTEM IN 5 SIMPLE STEPS

DANIEL I STEIN

RMC PUBLISHERS

© Copyright 2022 - All rights reserved.

It is not legal to reproduce, duplicate, or transmit any part of this document in either electronic means or in printed format. Recording of this publication is strictly prohibited and any storage of this document is not allowed unless with written permission from the publisher except for the use of brief quotations in a book review.

ALSO BY DANIEL I STEIN

Attracting Hummingbirds

The Essential Herb Gardening Handbook

Conquering Climate Change Anxiety

Check out my author page

CONTENTS

INTRODUCTION	1
Elements of a Rainwater Harvesting System	3
Common Rainwater Harvesting Questions	7
Rainwater Quality	8
STEP 1: PRE-PLANNING	11
Goals	13
Choices	18
Soft-Limits	20
Hard-Limits	22
Pre-Planning To-Do List	31
STEP 2: DESIGN	35
Collection Area	36
Pre-Filter	42
Transfer System	47
Storage Tank	58
Pump	70
Overflow	77
Make-up System	81
Treatment System	84
Connection System and Fixtures	92
Sensors & Controls	94
STEP 3: SELECTING COMPONENTS AND SAFETY CONCERNS	97
Selecting your RHS Components	98
Compatibility Checklist	108
Safety Concerns by RHS Component	110
STEP 4: INSTALLATION AND TESTING	117
Finding Professionals	118
Installation Checklist	119

Testing Checklist	124
STEP 5: MAINTENANCE	129
Maintenance Checklist	129
Cleaning Agents	134
Additional Events	135
Common Problems and Troubleshooting	136
EXAMPLE SYSTEMS	141
Basic Outdoor System	142
Large Outdoor System	148
Conclusion	155
Annual Rainfall by State	165
Additional Sources	167

FREE BONUS

Use my '**Outdoor Space Planner**' to:

- Get inspired, and choose what elements you *like*, *want*, or *need* to have in your new outdoor space
- Identify any special requirements or restrictions that might be roadblocks to improving your backyard
- Create an easy to follow budget so you don't run out of money, space, or time
- Draw out your personalized design and plan a schedule for everything to get delivered, built, and installed.

Go to https://www.subscribepage.com/danielistein to get it now!

INTRODUCTION

"When the well's dry, we know the worth of water."
Benjamin Franklin

We all have different reasons for wanting to harvest rainwater. Some of us are after sustainability and reducing our ecological footprint; others are trying to save money or become self-sufficient. And some of us want to be prepared for emergencies, with a secure water source for our family.

Whatever your motivation: sustainability, self-sufficiency, or security, having control over your water source is never the wrong decision.

This book will cover everything you need to design, install, and operate a rainwater harvesting system (RHS) for your home. There is a lot of flexibility in how you can design an

RHS, and they can fit almost any level of budget or expertise. Creating your RHS can range from a DIY weekend project to a significant home renovation, depending on your goals and how you want to use your water.

Once we cover some basics of what an RHS is and how they work, we'll go through the five steps to creating your very own rainwater harvesting system:

Step 1: Pre-planning
Step 2: Design
Step 3: Selecting Components
Step 4: Installation and Testing
Step 5: Maintenance

Not every part of each section will apply to your RHS. For example, suppose you do not want to use your rainwater as drinking water. In that case, you can skip over some of the treatment sections or requirements for indoor usage. Or, if you intend to use rainwater for basic landscaping, you may not need a pumping system or electrical hookups. These versions of an RHS are much simpler to design and install yourself.

Instructions for a basic outdoor system that anyone can install in an afternoon or two are included after Step 5. This example is a great place to start if you want a quick, reliable, and effective RHS without too much fuss. But, going through at least the first two pre-planning and design steps first will help you see what options are available and give you some ideas on how to tweak that basic RHS to meet your needs. By the end of this book, you will understand how an RHS is put together, works, and what maintenance they need.

Simple garden rain barrel

ELEMENTS OF A RAINWATER HARVESTING SYSTEM

First, let's go over the parts of a typical RHS. Of course, not every system will have all these parts, but we should understand what each part does and how they all work together. This information will allow us to design an RHS that achieves our goals. Parts below will be noted as either 'mandatory' or 'optional'.

If you see a term you don't recognize, check the glossary at the end of the book. There are a lot of specialized terms that might take a bit of time to become familiar with, and most of the parts of an RHS can have a few different names.

Collection Area (Mandatory), also known as a catchment area or collection surface

This is the surface that rainwater lands on and is harvested from. For most residential rainwater systems, this will be the roof of your home.

Connects to: Transfer System

Pre-Filter (Optional), also known as pre-treatment

Anything that removes debris or contamination from rainwater before it reaches the Storage Tank is considered pre-filtering. This is usually intended for physical contaminants and includes options like gutter guards and leaf filters.

Connects to: Transfer System

Transfer System (Mandatory), also known as a conveyance network or conveyance system

This is the system of gutters and pipes that moves the rainwater from the Collection Area to the Storage Tank. Some of this may already be present in your home, like the gutters on your roof, and some of it will need to be installed.

Connects to: Collection Area, Storage Tank

Storage Tank (Mandatory), also known as a basin, cistern, holding tank, water butt, or rain barrel

The Storage Tank holds the collected rainwater before it gets used. The Storage Tank is the central hub of the RHS. Everything comes into, sits in, or goes out of the Storage Tank.

Connects to: Transfer System, Connection System

Overflow (Mandatory), and Make-Up Systems (Optional)

A Storage Tank isn't just an empty tank; it may have some controls in place for extreme situations. Too much water (solved by an Overflow system) or too little water (solved by a Make-up system) can cause severe problems in your RHS. These systems aren't necessarily complex but have to be considered.

Connects to: Storage Tank

Pump (Optional)

Almost every system will have a pump unless the water is drained directly from an aboveground storage tank for nearby use. Pumps will sit inside, or near, the Storage Tank and usually require an electrical connection, but manual pumps are also an option. Additional pumps may be added after the Treatment System to reach fixtures further away or which are elevated above the Storage Tank.

Connects to: Storage tank

Connection System and Fixtures (Optional)

The Transfer System brings rainwater into your Storage Tank, and the Connection System moves that rainwater to where a fixture uses it. A fixture could be a tap, toilet, or appliance that uses rainwater.

Connects to: Storage tank

Treatment System (Optional), also known as post-treatment, filter, or disinfection

Most uses of rainwater will require some treatment, but there are a few cases where this isn't necessary. The level of treatment needed will be based on the amount of contamination picked up from the Collection Area surface and the required water quality for your intended uses.

Connects to: Connection System

Sensors and Controls (Optional),

Depending on the complexity of your system, you may need to monitor what is happening inside the Storage Tank. The most common combinations are water level sensors combined with valves. The sensors and controls could be linked to the Overflow, Make-up, or Pump Systems. However, you might also need sensors and controls for other parts of your RHS.

Connects to: Storage Tank

Cold Weather Modifications (Optional)

Several modifications can be made for an RHS operating outdoors in temperatures below freezing. These are specific to each element of the RHS and will be covered in Step 2.

Connects to: Various elements

COMMON RAINWATER HARVESTING QUESTIONS

Here is a compilation of the most commonly asked questions about rainwater harvesting:

Q: Is rainwater safe to drink?

A: Yes. Rainwater isn't pure but has much less contamination than most other sources, like groundwater or surface water. See 'Rainwater Quality' below for details.

Q: What skills do I need to install a rainwater system?

A: A basic rainwater harvesting system requires some basic plumbing skills, like being able to measure, cut, and fit pipes together. More complex designs will require some electrical work, engineering, and a basic understanding of water chemistry. However, many complex parts of an RHS, like Pumps or Treatment Systems, can be purchased as all-in-one packages that don't require extensive knowledge, tools, or skills.

Q: Does rainwater harvesting cause environmental damage?

A: Harvesting rainwater does not permanently remove water from the local environment. Instead, an RHS can help conserve local groundwater aquifers and slow down the effects of erosion from surface runoff.

Q: Is rainwater harvesting legal?

A: Rainwater harvesting itself is legal, but there may be restrictions or limitations on how you can use that water or how much you can collect. That will be covered in Step 1: Pre-planning but will require some additional research and double-checking based on your local laws and regulations.

RAINWATER QUALITY

Rainwater is nearly pure while it's in the air, but as soon as it hits a surface, there is a chance for it to become contaminated.

Water quality is essential if you plan to use your harvested rainwater as drinking water. But you will still want to maintain a high standard of water quality for other uses as well.

Contamination picked up from the Collection Area can make rainwater unsuitable for drinking, watering gardens, or other household uses.

There are three types of contamination to worry about: physical, chemical, and biological. These can all affect rainwater's taste, color, odor, and safety. These are discussed in detail in the section in Step 2 on Treatment Systems.

The best approach to maintaining the highest level of water quality is to use a multi-barrier approach. This approach is the gold standard used by any large-scale drinking water system. So while your RHS will be much smaller, the same principles can also benefit you. Using multiple methods to reduce contamination and keep rainwater clean through every step of the harvesting process ensures that we aren't relying on any single approach or piece of equipment to protect us. If one part of the Treatment System is not performing well, you will still receive high-quality water.

The RHS you design will incorporate a multi-barrier approach, and we will cover the details of water quality and treatment throughout the rest of the book.

STEP 1: PRE-PLANNING

This first step is to decide the purpose of your rainwater harvesting system (RHS). This decision will allow you to set limits for the design step that's coming next. Being clear on your **goals** and understanding what you want to achieve with your RHS will ensure your design is doing what you need it to do.

Many of the constraints on your RHS will be **choices** you make, for example, deciding what an acceptable budget is, or how much maintenance you can handle.

Other constraints will be **soft limits**. For example, the size of your roof will determine how much water you can collect, or your property's layout may limit the size of the Storage Tank you can install. These factors are challenging, but not impossible, to change.

And some constraints will be **hard limits**—things like your local climate or legal requirements that you cannot influence at all.

Pre-planning gives you a solid point to start from, but designing an RHS can be an iterative process. You might set goals or make choices in the pre-planning stage and then realize that you need to rethink things once you reach the design or component selection step. Maybe you will notice that you are running into some soft or hard limits or cannot achieve the goals you initially wanted. Perhaps there are limitations on what parts or contractors are available to you.

This back and forth is normal, and it is better to start pre-planning with an ideal system in mind and then see if you run into any difficulties in the design stage. At this point, you can make any necessary compromises to achieve as many of your goals as possible and revise your design accordingly. Of course, none of us has unlimited money or space, so there will always be a few compromises. However, most of us can still achieve our main harvesting goals with some creativity.

> *Important*: Don't get intimidated by this section. It might seem like this will be a massive project, but the effort it takes to create an RHS scales with the scope of your goals. For example, an RHS that supplements your existing water source and provides extra water for landscaping might cost a few hundred dollars with minimal planning and mainte-

STEP 1: PRE-PLANNING

nance. As you set goals and understand the limits of what you can do, some of the optional parts of an RHS or more complex decisions might become irrelevant. All that will be left are some straightforward choices. Then, if you decide to upgrade to a more complex RHS in the future you can come back to this section and plan a more ambitious RHS with more experience behind you.

GOALS

The first and most important question is 'what do you want to use your water for?'.

> "Less than 3% of treated water is used for drinking. The rest goes down the drain, down the toilet, or on our gardens."
> *Environment Canada*

Potable vs Non-potable

Potable means water specifically intended for human consumption. This isn't just water intended for use as drinking water but also includes any taps, sinks, or areas where food is prepared. You may also see the term *domestic water*, which is considered drinking water but includes other indoor uses like cleaning, bathing/showering, or activities where people could end up swallowing some of that water by accident.

Any water for potable or domestic use will have to meet the applicable drinking water standards set at the federal or state/provincial level. As a general rule, if you are connecting your RHS to fixtures inside your home, this water will likely have to meet drinking water standards. Therefore, you should consult a qualified professional when planning and constructing an RHS to supply potable water.

Some domestic uses have less strict quality requirements, often called gray water. Toilets and laundry are the most common fixtures which can use gray water, and do not need to meet drinking water quality standards.

Non-potable water includes any other water uses. This could be outdoor hoses for watering your lawn or garden, providing water for pets, or supplying water features. Some treatment might still be required for non-potable water, especially if it is used to provide water for animals or edible plants. The only exception is usually water intended for underground irrigation systems, which require no additional treatment. Still, you should confirm with your local regulations and laws.

Your first decision is whether you want potable or non-potable water. Then, list all the ways you might want to use your harvested rainwater. If your uses are primarily outdoors, you can design a reasonably simple system. If your primary uses are indoors, it will take additional planning to integrate the water from the RHS into your existing plumbing system. Common indoor non-potable uses are flushing toilets and

STEP 1: PRE-PLANNING

washing clothes (rainwater is naturally soft, making it ideal for washing).

Important: By default, any harvested rainwater is considered to be non-potable. It is strongly recommended that beginners start with an RHS intended only for non-potable outdoor applications. Non-potable uses make up the majority of most households' water, so these systems are still very effective.

Examples of non-potable outdoor water uses:

- Gardening and irrigation
- Filling ponds, pools, or hot tubs
- Water for wild animals
- Outdoor cleaning
- Washing vehicles
- Fire protection or emergency water

Important: water for pets and livestock doesn't necessarily have to be potable. But, you will likely want some amount of treatment to avoid unnecessary vet bills and harm to your furry and feathered friends.

How much water do you need?

Once you know how you would like to use your rainwater, you can figure out how much you need. Here is a quick cheat sheet to help you get a rough estimate of your water use, but you can also check directly on your fixtures for flow rates:

Indoor

Toilet (old): up to 5 gallons (22 L) per flush
Toilet (modern): 1.2 gallons (4.5 L) per flush
Laundry (top-loading): 30 gallons (140 L) per load
Laundry (front-loading): 20 gallons (90 L) per load
Bathroom sink: 1 gallon (4.5 L) per minute
Shower: 3 gallons (13.5 L) per minute
Kitchen sink: 2 gallons (9 L) per minute
Dishwasher: 6 gallons (27 L) per load

Outdoor

Garden hose (½ inch): 9 gallons (40 liters) per minute
Garden hose (⅝ inch): 17 gallons (77 liters) per minute
Garden hose (¾ inch): 23 gallons (105 liters) per minute
Irrigation system: 5.5 gallons (25 liters) for 1 inch (25 mm) of water per square yard (sq. m)
Garden: assume 0.6 gallons per sq. ft. of garden per week (25 L per sq. meter per week)

STEP 1: PRE-PLANNING

If you are stuck on this step and having trouble coming up with an estimate you feel confident in, keep a log of your water use over the next week or check your monthly bill.

Here are a few samples of generic monthly water budgets to check against:

Indoor use per person: 3,000 USG/month (11,300 L/month)
Assumes 100 UDG/day: 100 USG x 30 days

30 deck pots hand-watered: 240 USG/month (910 L/month)
Assumes 2 USG gallons per pot per week

Flower bed: 187 USG/month (710 L)
Assumes 25 feet long by 3 feet wide and 1 inch of water a week

Raised bed vegetable gardens: 60 USG/month (230 L)
Assumes 3ft by 8 ft and 1-inch of water per week

How will you use your RHS?

There is a big difference between trying to create a reliable primary water source and supplementing what you already have with rainwater as a backup system.

Suppose you want a consistent water supply that replaces your existing water source for particular uses. In that case, you will need to have a solid understanding of how much water you will be using and be able to capture and store enough water to meet those needs.

On the other hand, if you want your RHS to supplement your existing water supply, you can be a bit more flexible.

CHOICES

Your choices will influence how you will achieve your goals. These are entirely flexible and can easily be changed at any time in the design step.

If this is your first time building an RHS, you may have to return to this step a few times as you refine your design.

Budget

How much money can you invest in upfront costs for materials and installation?

How much can you afford in ongoing operating and maintenance costs? For example, if you are replacing existing water uses, you can subtract those costs from your RHS budget.

If you don't have the budget for an RHS that will meet all of your needs, you can design your system with expansion in mind. A good place to start is with half the capacity you need, and then plan the RHS layout to allow for future expansion.

STEP 1: PRE-PLANNING

Below Ground Installation

Would you be comfortable installing part of your RHS underground?

This option will require additional costs, maintenance, and complexity but allows for larger Storage Tanks and better use of limited space.

Maintenance

Every RHS requires maintenance; decide what a reasonable time is to spend on a weekly and annual basis.

Here are the following skills you may need for some maintenance tasks. Decide which ones you have and which you are willing to learn.

- Basic plumbing (checking seals for integrity, looking for clogs, inspecting pumps) - every RHS will need some basic plumbing skills
- Electrical (checking connections and that control units are operating) - if you have powered pumps or controls, you will need to be able to troubleshoot these systems
- Collecting water samples and interpreting basic water chemistry results - if you are using rainwater for drinking water, you will need to regularly test the water and compare the results against drinking water standards

- Water budgets (comparing expected vs. actual water volumes) - if you are relying on rainwater as your primary water source, you will need to spend some additional time confirming that you can collect enough water to meet your daily needs

SOFT-LIMITS

Collection Area

In most situations, the Collection Area will be your roof surface, and this will have a fixed area. There are ways to add additional space, for example, by combining a roof Collection Area with a detached garage or shed. However, the footprint of your home will be the primary factor in determining how much rainwater you can collect. Combined with your regional climate (see *Hard-Limits* below), this will determine the maximum quantity of water you can expect to collect.

The Collection Area is roughly equivalent to the square footage of your top floor. While it might be slightly larger due to some overhang, using that top floor square footage value will give you a conservative estimate that will take into account some losses and inefficiencies during collection.

> *Important*: when calculating your collection area, do not use the surface area of your roof unless it is completely flat; this will result in an overestimate.

STEP 1: PRE-PLANNING

Use the following formula to determine how much rainwater you can collect:

For every 100 square feet of indoor space on the top floor of your home, you can collect 62 US gallons (USG) for every inch of rainfall.

[sq. ft. of top floor] x [inches of rain] x 0.62 = USG collected

[sq. m of top floor] x [mm of rain] = L collected

This represents a Collection Area working perfectly, and we know that nothing in life ever works perfectly. We will take a more detailed look at how to factor in Collection Area inefficiencies in Step 2: Design, but this is a good start.

Also, make a note of your roof material at this time. The material will influence your collection efficiency and may affect the type of treatment required.

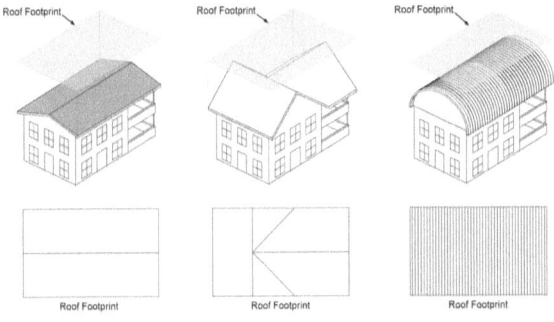

Determining Collection Area from roof size

Property Size

The layout and size of your property will limit where you can place a Storage Tank. While smaller aboveground tanks or rainwater barrels can fit almost anywhere, you may have fewer options if you want to use a larger tank. You may also need to have space between your Storage Tank and your building foundation, property line, and other structures.

Underground or buried tanks are limited by the location of buried utilities or service lines, which can be extremely costly to move.

Have a clear idea of the space available for aboveground tanks and where any utility or service lines are located that might prevent the placement of belowground tanks.

HARD-LIMITS

Climate - Precipitation

You can examine previous weather records to get a reasonable estimate of your expected rainfall in the future. Then, combined with the Collection Area calculation above, you can calculate the predicted volume of rainwater you can harvest. There are three numbers we will want to consider here.

The first is total precipitation over a year, in inches or mm.

STEP 1: PRE-PLANNING

The second is how many days of precipitation per year, or if there are any precipitation patterns between seasons. This information will confirm if you get steady rain which will keep refilling your Storage Tank as you use it, or if you can expect less frequent rains and need to ration what you collect.

The final number you will want is the maximum amount of rainwater you might receive at any one time, like a worst-case storm event. This value will be used to determine the size of your Transfer and Overflow Systems.

You might have a pretty good idea of what these numbers are already. Still, it helps to have actual values here to help decide what options best suit your climate before making any final decisions or investments.

In addition, there is quite a bit of variability between years, so if you can take average values from the last 5 or 10 years, you will get more realistic estimates for these numbers. Here are some places you can look to find this information:

United States of America

Go to https://www.weather.gov/

Click on your location on the map (you can hover over the map to see the location name). This will take you to your location home page, where you can select *Climate and Past Weather* and choose the *Local* option.

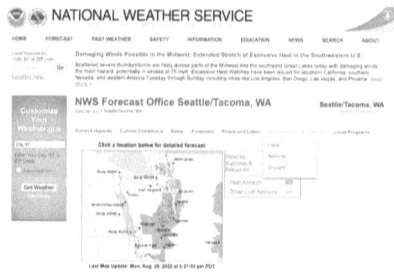

Under NOWData, you select the type of information you want. Start by picking the location closest to you in column 1. Then, under column 2 (Product), select accumulation graphs at the bottom. For column 3 Options, select a start date of January 1st for a historical year and an end date of December 31st of the previous year, and set the variable to Precipitation. Finally, click *Go* under column 4 View.

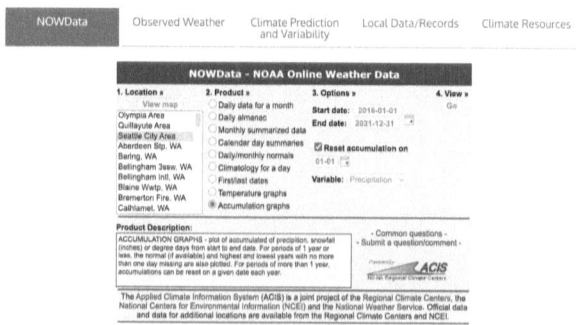

You will see a graph, and if you hover your cursor over December 31st on any year, there will be a pop-up with two important numbers.

STEP 1: PRE-PLANNING

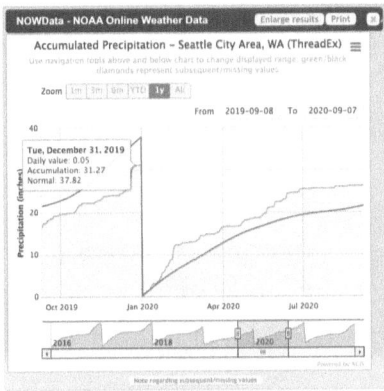

Accumulation is the total precipitation for that year, and *Normal* is the average annual precipitation from all the years you selected. The differences in *Accumulation* values will give a sense of the range of annual precipitation you can expect.

Another option is to select *Daily/Monthly Normals* under Product, and *Monthly* and *Precipitation* under Options. These options will give you the monthly precipitation values and then an annual normal. Unfortunately, you don't get the record low and high values this way.

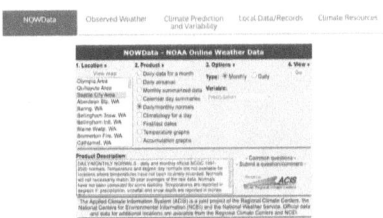

Next up is to check how many days there was precipitation. To find this number, select *Monthly summarized data* under

Product. Then, under options, put in the last 10 years as the date range, *Precipitation* as the variable, *Number of Days* under summary, and a threshold of > 0.

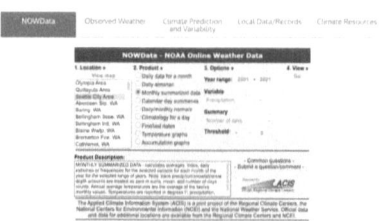

When you click View, you will see a large table. Scroll to the bottom, and look for the Mean, Max, and Min numbers on the far right. This is the average, highest, and lowest number of days of precipitation you can expect over the year.

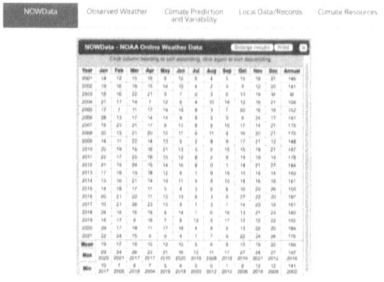

This website, weather.gov, is not the only way to get this information. If you are familiar with other weather tracking websites or used to using other sources, that won't be a problem. Another alternative you can also check is the FMP Rainwater Harvesting Tool.

STEP 1: PRE-PLANNING

Canada

Go to https://climate.weather.gc.ca/climate_normals/

Choose *Search by Proximity*, enter your city, select *Go*, pick the weather station closest to you, and record the following: Yearly rainfall (mm) and the largest extreme daily rainfall (mm).

This selection will give you an approximate total amount of water you can harvest over a year and the most extreme rain event you can expect.

Europe

Go to https://www.currentresults.com/Weather/Europe/Cities/precipitation-annual-average.php

This will give you a good starting point for total precipitation and number of days of rain. Check with your country's environment or weather departments for more specific data for your location.

> *Important*: You must also adjust for losses and any inefficiencies during collection. We will cover that in detail in Step 2: Design, but for now, assume that only about 80% of the rain landing on your Collection Area will make it to your Storage Tank.

Climate - Freezing Temperatures

Does the temperature drop below freezing at any time of the year? If yes, you may have to include some cold-weather modifications. These modifications are marked throughout the remaining steps and may add additional cost and complexity to your RHS.

You may also have to adjust your precipitation estimates for snow if you regularly clear off your roof during winter, preventing snow from reaching your Storage Tank after melting.

Local Laws and Regulations

The most important consideration when designing and installing an RWH system is the relevant codes, regulations, standards, and laws. These are the legal frameworks you will have to work within, no matter how outdated or illogical they might be. Other rules, like strata or homeowner association bylaws or community-specific concerns, will also have to be considered.

Currently, most countries do not have consistent federal laws or regulations on collecting, storing, or using rainwater safely.

> *Important*: If you choose to ignore these hard limits, you should at least understand why they are in place first. It is one thing to willfully ignore a minor bylaw that is out-of-date and inconsistent with current environmental practices.

STEP 1: PRE-PLANNING

But it is another to ignore regulations intended to protect your family's health and safety.

The most common restriction is against using rainwater for drinking or domestic purposes. However, there may also be restrictions on the volume you can collect, placement of Storage Tanks, or materials you can use in the Transfer or Connection Systems.

Here are a few ways you can confirm what your local regulations and laws are:

- Ask your state or provincial departments; this could fall under many different departments, such as environment, health, agriculture, natural resources, housing, or water boards.
- Ask your municipality who you can contact to find out about local bylaws or restrictions. These often appear in residential housing, plumbing codes, or sewerage regulations. Building inspectors are also a good resource.
- Search for rainwater harvesting or collection materials produced by local or state/provincial agencies. While some areas may have restrictions, drier climates may promote this practice and offer incentives and rebates.

Applicable regulations could include:

- Plumbing Codes
- Building Codes
- Electrical Codes
- Certification for tanks, pipes, and treatment
- Water quality standards or guidelines

A summary of rainwater harvesting regulations and links to state specific information across the USA can be found here:

https://www.energy.gov/eere/femp/rainwater-harvesting-regulations-map

PRE-PLANNING TO-DO LIST

Make sure you have answers to the following questions:

Do I want potable or non-potable water?
What am I planning on using harvested rainwater for?
How much water do I need per month for these uses?
Will rainwater be the primary or backup source?

What is my budget to purchase and install the RHS?
What is my monthly budget to operate the RHS?
Would I consider installing a below-ground Storage Tank?
How much maintenance do I want regularly and annually?

How large is my Collection Area (sq. ft. or sq. m)
What type of material is my collection area covered with?
How much space do I have for a Storage Tank (sq. ft. or sq. m)
What are my preferred spots for placing a Storage Tank?

What is the average yearly precipitation (inches or mm)?
What is the low-end of yearly precipitation (inches or mm)?
What is the high-end of yearly precipitation (inches or mm)?
How many days of rain can I expect per year?
Which months have the most frequent rain?
What is the biggest rain event in recent history?
Will my RHS need cold-weather modifications?

What local laws or restrictions are in place for collecting or using rainwater?

Which of the following skills do I have or could learn:

- Basic plumbing
- Basic electrical
- Interpreting water chemistry results
- Creating a monthly water budget

STEP 2: DESIGN

This step will go through every component of an RHS. We will cover how to select each part and how to ensure they are compatible with each other. Most of the design choices will be made based on your pre-planning, and not every section of the step will be relevant to your RHS.

This step is the largest in creating your RHS, and you may have to refer back to this step later in the process. However, a well-designed system will be easier to purchase, install, and maintain, so extra time and effort spent here will pay off in the end.

Step 2 is probably the most challenging part of the process and involves several decisions which might be interdependent. Suppose it seems like the RHS is getting too elaborate. In that case, you can always go back to pre-planning and realign your goals with the level of complexity you are comfortable dealing with. Also, remember that if your RHS doesn't need a specific component, you can skip over that section for now.

Another approach is to design your RHS, knowing that you will hire professional help to confirm your decisions and help with installation. If you know what you want, how it all works together, and the right questions to ask, you don't necessarily have to select every part and install the entire RHS yourself.

> *Important*: Cold weather modifications are included individually for each RHS component.

COLLECTION AREA

As previously discussed, your roof is the most common and best option for a Collection Area. Also referred to as roof catchments, this is almost always the best option because it is almost always the largest surface you have and will give you the largest amount of harvested rainwater. However, if you have minimal water needs, even a tiny garden shed could collect upwards of 500 USG per year.

You can explore other options for Collection Areas, such as lawns, driveways, or paved areas, but there are two main drawbacks. First, water quality from these surfaces is generally lower quality and requires additional treatment. Second, since roofs are higher, they can use gravity to move water through the Transfer System to the Storage Tank. If you choose to use a ground catchment or alternative surface, you may need to consult a qualified professional to support the design and installation of a more complex RHS. These designs will also require active pumping to move water back through your Transfer System to the Storage Tank.

Overflowing rain gutter

The design considerations here apply to most types of Collection Areas. Still, we will assume going forward that you will be relying on a roof catchment. This option means there won't be additional materials, installation, or maintenance costs.

It is possible to make changes to your roof to support an RHS, but this is expensive and will not be cost-effective. Therefore, this option is not recommended unless you plan to do a significant renovation or replacement work on your roof.

Another option is to link multiple roofs together and direct them to the same Storage Tank. So, if you currently have multiple structures close to each other on your property, this would be the easiest way to increase your Collection Area.

Materials

The material covering your roof affects the quality of the rainwater collected and the efficiency of collection. The material will also influence the type of pre-filtering, treatment, and maintenance needed.

Water Quality

There are two ways that the Collection Area influences water quality. First, debris on the Collection Area can introduce contamination into the rainwater. Debris could include dust, foliage, bird or animal droppings, and anything blown or dropped onto the roof surface. The second way is through roof materials releasing chemicals or tiny particles directly into the water.

Dust and debris can be addressed by keeping the Collection Area clean and removing sources of leaves and twigs like overhanging branches. Pre-filtering systems can remove some of the larger debris as well.

For contamination directly from the roof material itself, here is a rough classification of how roof materials affect rainwater quality:

STEP 2: DESIGN

- *Materials that can result in poor quality water*: treated wood (chemically treated and poor efficiency), galvanized metal (often coated with toxic cadmium), lead flashing (contains lead), asphalt shingles (contains toxins and releases fine particles)
- *Materials that have minimal impact on water quality*: epoxy or enamel-coated metal, fiberglass, galvalume metal, EPDM (Ethylene Propylene Diene Monomer) without fire retardants, clay tile, ceramic, terracotta, and solar panels
- *Materials that do not introduce contaminants above drinking water guidelines*: materials meeting the NSF/ANSI 61 Standard (for potable water) or the NSF P151 Standard (this one is specific to rainwater harvesting). Some materials may have NSF 61 coatings applied and would be suitable for a Collection Area.

Important: Asphalt or fiberglass roofs may be coated with anti-fungal chemicals to prevent moss growth. These chemicals may make water unsuitable for drinking without advanced treatment methods, and can also harm some plants. Therefore, designing an RHS with a Collection Area made with either of these materials is not recommended.

Efficiency

The ability of your Collection Area to capture rainwater is called the collection efficiency. It is the proportion of water collected compared to the total amount of rainfall.

There are two types of inefficiencies to consider for Collection Areas. The first is loss due to the initial rain absorbed into the roof material. For example, the first 0.02 inch (0.5 mm) of rain could be absorbed by the roof surface and never reach the Transfer System before evaporating.

The second is described as a percentage loss of the continued rain afterward. As a rule of thumb, we can expect a 20% loss this way. You can get a rough idea of your collection efficiency by applying the factors below:

Surface efficiency
Metal = 95%
Shingled = 90%
Terracotta, clay, or tiles = 85%
Wood = 80%

The initial loss calculation is difficult to factor accurately, so we will only use the overall surface efficiency. When calculating the total volume of rain we can expect to collect, here is how we can account for collection efficiency:

STEP 2: DESIGN

[sq. ft of top floor] x [inches of annual precipitation] x [surface efficiency] x 0.62 = expected annual harvested water in USG

[sq. m of top floor] x [mm of annual precipitation] x [surface efficiency] = expected annual harvested water in L

As an example, if we are expecting on average 40 inches of rain per year on a 1,000 square foot metal roof, we would expect to collect:

1,000 sq. ft x 40 inches x 0.95 x 0.62 = 23,560 USG annually

Calculation Shortcut

For every 100 sq. ft of Collection Area and 1 inch of rain received, you can fill one standard sized 55 USG barrel.

Snow Rails

If you have a sloped metal roof, there may be snow rails. These prevent damage to gutter systems by slowing down melting snow as it slides off the roof. Snow rails will also benefit your RHS as they hold snow rather than let it fall off the roof. This snow will eventually get into the Transfer System as the temperature warms.

Other Considerations

Depending on how you set up your Transfer System, your RHS may capture only a portion of your Collection Area. For example, if each side of your home drains into different downspouts. Ensure that any estimates of how much water you will be harvesting consider this. Suppose only 50% of your roof area drains into your Transfer System. In that case, you will need to adjust the size of your Collection Area accordingly.

PRE-FILTER

The purpose of a Pre-Filter (or pre-treatment system) is to reduce damage and wear on other components of the RHS, especially the Treatment System. Any modification to the Collection Area or Transfer System intended to prevent contamination from reaching the Storage Tank will be considered a form of pre-filter.

Pre-Filters are not as effective at removing contamination as Treatment Systems. Still, they are an important part of the multi-barrier approach. A well-designed Pre-Filter can significantly improve the quality of water entering the Storage Tank, reducing the workload and increasing the effectiveness of the Treatment System. In addition, pre-filters reduce overall maintenance costs, as Treatment Systems are often the most complex, challenging to access, and most expensive part of your RHS.

Here are the most common types of Pre-Filter:

- First-flush diversion - an initial amount of collected rainwater is diverted away from the Storage Tank by

STEP 2: DESIGN

modifying the Transfer System. The first portion of rain will contain the most debris and contamination, but this diversion will also reduce collection efficiency by about 20%. First-flush diversion is a low-tech and effective option if you receive more rainfall than you need to meet your water needs, and is the most effective way to improve the quality of your collected rainwater. You can calculate the size of the first-flush diversion you need using the formula below.

- Screens or filters can be added to prevent physical contaminants from entering the Transfer System. These options require regular maintenance to prevent clogs or reductions in Transfer System flow, but they are much easier to reach than clogs within the Transfer System pipes. These are also easy to install and could include gutter guards and leaf screens over downspouts. Or, screens and filters can also be integrated right into the pipes of a Transfer System. There is no reason not to include some form of screen in your RHS.
- Settling tank - This is a slightly more complex option. Before rainwater is delivered to the primary Storage Tank, water is first conveyed to a settling tank, where any contaminants in the water can sink to the bottom. This settling tank is usually directly connected or integrated into the Storage Tank. As a

general rule-of-thumb, a settling tank will be approximately one-third of the volume of your Storage Tank. There isn't any loss in collection efficiency, as this water will eventually reach the Storage Tank. Settling tanks can be an extremely effective way to remove physical contamination like dirt, dust, and physical debris, but they also increase the complexity, cost, and maintenance needs of your Storage Tank.

Ideally, you would use all three Pre-Filter methods. But, if you are trying to minimize losses and maximize collection efficiency, you can use the following rule. If you get shorter, lighter rain events, use filters and screens. If you get heavier, longer rain events, use a first-flush diversion.

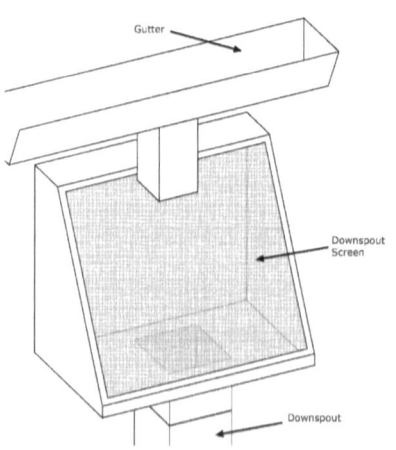

Pre-filter placed before a downspout

STEP 2: DESIGN

First-Flush Diversion

A simple design for a first-flush diversion is to add a vertical section of pipe that exits from the bottom of a horizontal section of the Transfer System, with a dripper valve on the bottom. After the vertical diversion section fills up, rainwater can continue normally through the Transfer System. The water stored in the vertical section will slowly drain over a few days, and the first-flush diversion will be ready for the next rain event.

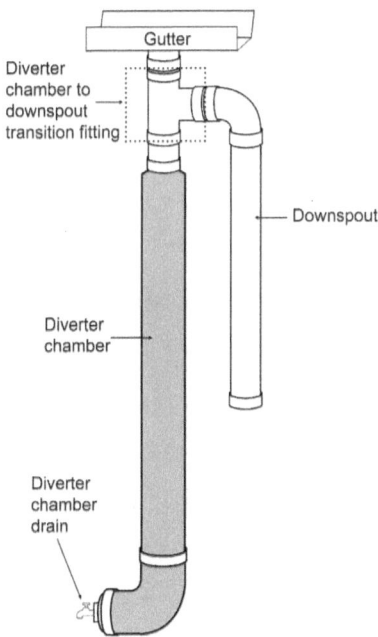

Simple first-flush system

Important: If math is not your strong suit, ignore the calculation section below. Just assume that for every 1,000 sq. ft of your Collection Area, you will need approximately 15 feet of 4-inch pipe in your first-flush diversion.

Typical first-flush diversions remove the first 0.04 inches (1 mm) of rainfall. Starting with that assumption, you can calculate the height of a first-flush diversion as follows:

Diversion Volume (cubic feet) =
0.04 inches x 0.083 ft/inch x Collection Area (sq. ft)
Height of first-flush diversion (feet) =
Diversion Volume / [3.14 x Pipe Diameter (inches)]^2

Diversion Volume (L) = 1 mm x Collection Area (m^2)
Height of first-flush diversion (m) =
Diversion Volume (L) x 1000 / [3.14 x Pipe Diameter (mm)]^2

For example, if you want to add a first-flush diversion and you have a 1,000 sq ft Collection Area with a 12-inch diameter diversion chamber. Your first-flush diversion chamber would have a minimum height of:

Diversion Volume (cubic feet) =
0.04 inches x 0.083 ft/inch x 1,000 sq ft = 3.3 cubic feet

STEP 2: DESIGN

Height of first-flush diversion (feet) =
3.3 cubic feet / [3.14 x (0.5 feet)^2] = 4.2 feet

You can size the first-flush diversion to meet your needs or the expected level of contamination that accumulates on the Collection Area. For example, removing the first 0.04 inches (1 mm) of rain is more than enough for even a highly contaminated roof. If you have a roof made from high-quality material kept relatively clean, diverting only the first 0.02 inches (0.5 mm) may be sufficient, so you can cut the first-flush diversion length in half.

Cold Weather Modifications

Pre-Filter components are outdoors and are susceptible to cold or freezing conditions. Check for ice build-up on filters or screens and remove any organic debris, like leaves, more frequently in freezing temperatures.

Cold weather modifications for first-flush diversions and settling tanks can use the same options discussed in the sections below for Transfer Systems and Storage Tank, respectively.

TRANSFER SYSTEM

The Transfer System of an RHS system typically consists of three main parts: external gutters (or eavestroughs), downspouts, and drainage piping (or rain leaders).

These three elements are usually already present on residential roofs, but sheds or other structures may not have them. Some modification to these existing elements is typically necessary to use them for rainwater harvesting. These include:

- Ensuring all parts are completely sealed from pests, animals, and debris
- Installation of additional pre-filters to maintain water quality
- Addition of a secure connection to the Storage Tank

Two main factors dictate the overall complexity of the Transfer System:

The first factor is the distance between the Collection Area and the Storage Tank. This distance can be minimal, just a downspout going directly vertical from the eavestrough into an above-ground tank. Or it can be longer and more complex, with linked Collection Areas or buried drainage piping across a yard connecting to an underground Storage Tank. The best practice is to try and limit the distance between the Collection Area and Storage Tank as much as possible.

The second factor is the worst-case conditions. What is the largest rain event your Transfer System needs to be able to handle? Ensuring the downspouts and drainage piping don't

STEP 2: DESIGN

overflow or backup during storm events will prevent your RHS from being damaged during storms.

Two critical design decisions can address this factor: pipe size and pipe slope. Proper sizing and slope of the Transfer System will ensure that rainwater is rapidly conveyed to the Storage Tank.

Pipe Size

Gutters

The gutter size should be based on the Collection Area and the number of downspouts that drain the water away from the roof surface. There usually aren't requirements or codes for specific gutter and downspout sizing; standard sizes and general approximations are traditionally applied.

A standard 5-inch (125 mm) K-style gutter is the most common. They are suitable for almost all residential-style roofs, drainage areas, and gutter lengths.

Suppose you need to add gutters to your Collection Area. In that case, you can calculate the required gutter size using the following method:

1. Calculate the size of the drainage area supplying the gutter as follows:

[length of gutter draining into downspout] x [depth of the roof]

2. Use the following list and pick an appropriate size based on the maximum expected rainfall intensity over 15 minutes:

- Rainfall intensity less than 1.4 inches (35 mm); use a 4-inch (100 mm) K-Style gutter on any size roof.
- Rainfall intensity less than 2.8 inches (70 mm) and drainage area less than 375 sq. ft (35 sq. m); use a 4-inch (100 mm) K-Style gutter.
- Rainfall intensity less than 2.8 inches (70 mm) and drainage area greater than 375 sq. ft (35 sq. m); use a 5-inch (125 mm) K-Style gutter.
- Rainfall intensity less than 5.1 inches (130 mm) and drainage area less than 375 sq. ft (35 sq. m), use a 5-inch (125 mm) K-Style gutter.
- If rainfall intensity is greater than 5.1 inches (130 mm) or greater than 2.8 inches (70 mm) with a drainage area over 375 sq. ft (35 sq. m); use a 6-inch (150 mm) K-Style gutter

Downspouts

The downspout size will be based on your selected gutter size. Generally, 2 x 3 inch (50 x 75 mm) rectangular downspouts or 3 x 3 inch (75 x 75 mm) square downspouts are suitable for most gutter sizes.

You can select a specific downspout size using the following rules:

STEP 2: DESIGN

- 4 inch (100 mm) K-Style gutter - 2x3 inch (50x75 mm) or 3x3 inch (75x75 mm) downspout
- 5 inch (125 mm) K-Style gutter - 2x3 inch (50x75 mm) or 3x3 inch (75x75 mm) downspout
- 6 inch (150 mm) K-Style gutter - 3x4 inch (75x100 mm) or 4x4 inch (100x100 mm) downspout

Drainage and Additional Piping

If you need additional piping in your Transfer System, it should have the same size or larger cross-sectional area than the recommended downspout.

Here is a quick conversion from square downspouts to minimum equivalent circular piping sizes:

- 4 inch (100 mm) square gutter = 16 sq. inch (100 sq. cm), circular equivalent pipe diameter of 4.5 inches (11.4 cm)
- 5 inch (125 mm) square gutter = 25 sq. inch (160 sq. cm) - circular pipe diameter of 5.6 inches (14.2 cm)
- 6 inch (150 mm) square gutter = 36 sq. inch (230 sq. cm) - circular pipe diameter of 6.8 inches (17.3 cm)

Important: Every part of the Transfer System must be sized to handle runoff from the Collection Area in extreme weather, with sufficient slope to promote rapid drainage. This will prevent the RHS from backing up and flooding.

Pipe Slope

The Transfer System works with gravity, so every section should slope toward the Storage Tank. A minimum slope of 0.5% should be maintained at all times, with a target of 2% throughout the Transfer System. A 2% slope is also equivalent to 1/16" drop per linear foot. The closer to vertical you can get, the better. You can calculate the slope percentage as follows:

[Vertical Distance] / [Horizontal Distance] x 100

This same requirement holds for buried and indoor pipes and will ensure that the Transfer System drains completely between rain events and remains dry most of the time. Standing water in your Transfer System can lead to problems such as insects, algae, and mold growth.

Transitions

Depending on the type of gutters and downspouts and the type of Transfer System pipes selected, you may need to find transition fittings that can connect the two systems.

Most common gutter or downspout and pipe types will have readily available adaptors; however, if you cannot find an adapter, consult a plumber to discuss potential options.

Diversion

Every Transfer System should have some method to divert collected rainwater from the Storage Tank if necessary. Diver-

STEP 2: DESIGN

sion could be for temporary events like cleaning the Collection Area or gutters or for seasonal shutdowns during the winter.

The simplest way to add a diversion to the Transfer System is with a single valve that redirects rainwater to the original downspout and closes off the entrance to the Transfer System.

Dry or Wet Systems

A dry Transfer System uses gravity to move water into the Storage Tank. It is completely emptied after the rain event. A wet Transfer System requires a pump or additional tank to hold collected water before it is moved against gravity into the Storage Tank. While dry systems are much easier to design, build, and maintain, a wet system may be required if:

- you need to move rainwater uphill,
- if you are using an elevated or rooftop Storage Tank, or
- if you need to run a portion of the Transfer System underground and connect to an above-ground Storage Tank.

Wet systems should only be used as a last resort and be designed with the support of a qualified professional.

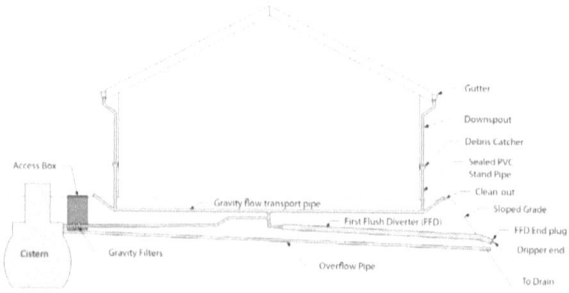

Example of an underground dry Transfer System

Planning the Transfer System Layout

Your goal is to have the shortest distance between the Collection Area downspout and the Storage Tank. The most straightforward Transfer Systems have the gutter drain vertically through a downspout directly into the Storage Tank.

First, determine where the primary downspout and the Storage Tank will be. In general, the Storage Tank location is more flexible. If there are multiple downspouts on your Collection Area, your options are to:

- Only use one downspout and have a smaller Collection Area, based on what portion of the roof catchment drains through that downspout;
- Connect multiple downspouts to your Storage Tank through a more complex Transfer System; or
- Remove extra downspouts, and increase your gutter size accordingly.

STEP 2: DESIGN

Important: If you are modifying or adding a gutter system, remember that downspouts should not be on inside corners or drain more than 50 feet (15 m) of gutter length.

Once you have the locations of your Storage Tank and Downspouts fixed, you can determine the best path between the two. A straight line is the best option; you should aim to minimize bends or corners in your Transfer System. First, measure the height of your Collection Area and the horizontal distance between the start of your Transfer System and your Storage Tank location. Then determine the maximum height of your Storage Tank, assuming a 2% slope.

Example:

Distance to Storage Tank of 10 feet (120 inches)
[120 inches distance] x [2% slope] = 2.4 inches drop

Gutter height of 12 feet (144 inches)
Maximum height of Storage Tank = 144 - 2.4 = 141.6 inches

In this case, there may be plenty of opportunity to have a larger slope, or to use a fairly tall Storage Tank.

Important: If there is a horizontal distance greater than 10 feet (3 m) between the Collection Area and Storage Tank, then the Transfer System will need additional supports.

Buried Lines

Below-ground pipes should be placed in a properly excavated space, supported, and properly backfilled according to the manufacturer's specifications and local building codes. Before excavating, check for buried utility or service lines, including gas, water, electricity, phone, or internet.

You can receive this information by contacting your municipality and any service providers. Many places have a *call before you dig* hotline or *click before you dig* website.

When planning the layout of buried lines, remember the initial and final burial depths will be different to account for a minimum 2% downwards slope.

If possible, ensure that buried pipes are located below the frost penetration depth. This depth will vary based on location, soil type, and local climate. Your local first penetration depth may be written into your local building codes or can be looked up online. If this isn't possible to install below frost penetration depth or the pipes are at ground level (not recommended), these pipes should have insulation or heat tracing.

Underground non-metallic pipes should be installed with 'tracer tape' (also referred to as 'tracer wire') 12 inches (300 mm) above the line. This is so you can easily find these pipes for maintenance, repair, or if you need to do digging or soil disturbance in the vicinity of your Transfer System.

STEP 2: DESIGN

Cold Weather Modifications

Rainwater can freeze in your Transfer System if not drained or temperature controlled. In addition, any blockages in the Transfer System increase the likelihood of freezing, which can severely damage pipes. Two main options for dealing with freezing temperatures are disconnecting the RHS or winterizing it.

Frozen downspout pipes

Disconnecting during Winter

This option diverts water away from the Storage Tank during the winter months. The Transfer System is either disconnected entirely (returning to your original gutters and downspouts), or a bypass is used to reroute water away from the Storage Tank. You must check that no water has pooled or remains within the Transfer System before temperatures reach below freezing.

This is the cheapest and easiest option if your usage is mostly in summer and you don't need to collect water during the winter.

Winterizing

There are two main ways to keep your Transfer System temperature above 32°F (0°C). First, remember that all outdoor RHS components, not just the Transfer System, may need cold weather modifications.

- Insulation: if you only have freezing temperatures intermittently, you might be able to use rigid Styrofoam insulation around your Transfer System pipes.
- Heating: adding heat trace wire around gutters and downspouts can keep water warm enough to reach the Storage Tank without freezing. Heating these areas instead of Transfer System pipes will avoid many problems with maintenance and running long lengths of heat trace wire.

STORAGE TANK

There are two main design questions for Storage Tanks: What size tank do I need? And where should I put it?

The Storage Tank is the central hub of your RHS, and many of the decisions for other components will be based on the

STEP 2: DESIGN

size and placement of the Storage Tank.

A well-designed RHS will have the smallest Storage Tank possible that meets your needs. However, there may be several additional components of the RHS that are installed within the Storage Tank.

These are covered separately, but your Storage Tank will have to be large enough to accommodate any of these extra components your design requires.

Size

Storage tank volumes can range from 40 USG (180 L) for a small rain barrel to thousands of gallons (tens of thousands of liters) for commercial-sized holding tanks.

If your tank is too small, it may overflow during rain events reducing your collection efficiency. However, some amount of overflow is acceptable if you are still able to meet your water needs.

If your tank is too large, it will cost more to purchase, install, and maintain and may unnecessarily limit your options for placing the tank.

A standard 55 USG (210 L) rain barrel is an excellent place to start if you want seasonal outdoor use. A 2,500 USG (10,000 L) Storage Tank for indoor use will be more than sufficient for most homes' daily needs. A large house with lots of landscaping might need a Storage Tank in the 10,000 USG (40,000

L) range; however, an RHS of this size would need support from qualified professionals for design and installation.

The maximum size Storage Tank you need will be based on how much water you expect to collect (see previous Collection Area section) and the amount of water you will be using based on your goals in Step 1: Pre-Planning.

> *Important*: Overall capacity of your RHS does not increase directly with the size of your Storage Tank. Small tanks which get used frequently are regularly emptied and refilled. You only lose out on capacity when a small tank overflows. If you regularly use your collected rainwater, you can get by with a much smaller Storage Tank than you might think.

Consider the following and how it applies to where you live:

- Do you get regular rainfall, or will most of it happen within a certain period? If you get steady rain year-round, you can divide your estimated annual collection volume by 12 to get your rough input into your Storage Tank. If you get most of your rain in certain months, you can divide your total expected collection volume by the number of months with heavy rain. You can check precipitation patterns for your area to confirm if you will have water when you most need it.

STEP 2: DESIGN

- Compare your monthly water budget for your heaviest use months against your expected monthly collection volume. If your budget is lower than what you could be collecting, then you should have enough water to meet all your needs and your Storage Tank can be sized based on your water budget. If your water budget is higher, you should select a tank large enough to store all the water your Collection Area can receive. You may also want to revise your water budget accordingly.
- This may take some refinement. For example, if you primarily want water for gardening or landscaping and will only have high demand during a few summer months. You can build up your bank of rainwater over the rest of the year. In that case, you may wish to select a Storage Tank large enough to capture your total annual volume of rainwater harvested by your Collection Area.
- Finally, multiply that desired Storage Tank size by 1.2 to account for 20% of the Storage Tank being unusable dead space. Dead space includes a section at the bottom below any Connection pipes, a section at the top above the Overflow, and any space used up by Treatment Systems, Sensors, or Controls installed within the Storage Tank.

Most rain barrels have an approximately 2 sq. ft (60 sq. cm) footprint. Larger Storage Tanks come in a variety of different shapes with different options for footprints, such as circular (most efficient use of space) or thin (can stand next to buildings).

You can consult a water balance table to help determine the optimal, most cost-effective Storage Tank size. A water balance table compares rainwater supply against total water demand to determine the water volume left in the storage tank at the end of each month. This calculation helps to estimate what size of the tank would provide for projected demands without running dry during the summer.

Placement

It is vital to ensure that your Storage Tank is in a stable and easily accessible location. While the placement of the Storage Tank will also affect the complexity and size of the Transfer System, stability and accessibility should take priority.

The slope, obstacles, and layout of your property will determine potential spots for the Storage Tank, but first, you need to make a few decisions:

Above-ground, below-ground, or integrated?

There are pros and cons to each option:

Above-ground is the most straightforward and cheapest option since there are no excavation costs. But you may have

STEP 2: DESIGN

to add additional cold weather modifications to protect against freezing temperatures, and large Storage Tanks can take up a considerable amount of space. Above-ground installation is recommended for smaller Storage Tanks (like rain barrels) and warm weather climates where freezing is not a concern.

Above-ground Storage Tanks should not be placed in areas where direct sunlight can reach inside the tank.

Preformed concrete below-ground Storage Tank

Below-ground Storage Tanks are more expensive and require additional planning to find an appropriate location to avoid buried utilities or service lines. But you can avoid cold weather modifications if they are buried beneath the frost penetration depth. Buried tanks also don't use up any space in your yard or on your property.

Below-ground installation is recommended for larger Storage Tanks or smaller areas or if you can easily install the Storage Tank below frost penetration depth. There may be building codes, regulations, or permits needed to bury a Storage Tank. Ensure the location is accessible for the initial excavation (ask a contractor), tank delivery (confirm with the supplier), and maintenance.

Integrated Storage Tanks are built directly into other structures, like your home, and are often directly incorporated into the building foundation. These are custom solutions, very complex, and expensive. This option is usually only necessary if you use rainwater as your primary drinking water source and below-ground tanks are not an option.

Integrated Storage Tanks won't be covered in detail in this book. They will need significant support from engineers and contractors and are not DIY projects. Please consult an engineer or architect if you are looking to design an integrated Storage Tank.

Once you have decided on a type of Storage Tank (above or below ground), you can decide on some potential locations that meet the following requirements:

- the Transfer System can quickly drain into the Storage Tank with a sufficient gradient (at least 0.5%, ideally 2% or more)
- the Overflow will be able to drain into a discharge area properly and away from other structures
- the ground is flat, stable, and well packed

STEP 2: DESIGN

- there is enough room to allow access for regular maintenance and inspection
- meets any local building code or regulatory requirements for separation from buildings, property lines, or other structures

Once you have a few potential locations, you can select between them based on your priorities; consider things like:

- reducing the length of your Transfer and Connection Systems as much as possible
- avoiding other uses of your space or inconvenient areas
- avoiding property lines, building foundations, or other permanent structures

Elevation

For simple above-ground Storage Tanks, elevating the tank will create some water pressure for free without having to install a pump. Sitting your Storage Tank on an elevated platform made of wood, gravel and cinder blocks, or concrete are all straightforward options.

Elevating your Storage Tank can give you an extra 4 PSI of pressure for every 10 feet (3 m) of elevation.

Check with your local codes and regulations for restrictions on how rain barrels or Storage Tanks can be placed or requirements for designing a platform to elevate your system.

Small Storage Tanks on elevated platforms

Materials

Now that the main questions have been answered, we can address some secondary details. There are several different types of tanks that can be made from different materials. You can now choose the best option for your RHS.

Local building codes or regulations may require the use of specific materials based on Storage Tank size, placement, (indoor vs. outdoor, above ground vs. below ground), and use case (domestic vs. non-potable). For example, suppose you are using harvested rainwater for domestic purposes. In that case, you will need to ensure your Storage Tank meets NSF/ANSI Standard 61 to ensure that contaminants are not introduced into the water.

The most common materials for Storage Tanks are concrete, plastic, and fiberglass.

STEP 2: DESIGN

Plastic - High-density polyethylene (HDPE) is a common choice for Storage Tanks under 5,000 USG (20,000 L). They are durable and light enough to transport and position easily.

Metal - For Storage Tanks over 5,000 USG (20,000 L), metal tanks are recommended.

Concrete - for permanent Storage Tanks, especially if installed below-ground, concrete may be a more practical option. If you do not have space or access for a larger HDPE or metal tank to be delivered, pouring a concrete Storage Tank may be the best alternative.

Natural - While we haven't covered using open natural storage areas, creating ponds lined with impermeable geo-membranes can be an option for some applications. This option essentially creates an artificial water feature using rainwater. It will require some serious planning and potentially permitting as well.

Preparing Soil for an Above Ground Tank

If the Storage Tank is being placed directly onto the soil, prepare the area as follows:

- Dig down at least 1 foot (30 cm), and compact the soil beneath
- Cover with a level layer of sand or fine gravel back up to the ground surface

- Top with concrete or stone slabs for the Storage Tank to sit on
- For larger Storage Tanks, follow manufacturer's instructions

Additional Considerations for Buried Tanks

- Even if you are not burying any pipe, you must ensure there are no utilities or buried lines in the vicinity of an underground tank.
- The maximum depth will depend on the material and the manufacturer's specifications. If you are burning your Storage Tank specifically to avoid cold weather modifications, you will need to select a material rated for a depth greater than the frost penetration depth in your area.
- If you are burying your Transfer System, ensure that the Storage Tank will be deep enough to allow for a minimum 2% grade drop of the elevation system.
- A typical Storage Tank used by a single family does not require any venting, as the Transfer System and Overflow drainage allow enough air exchange. However, larger tanks should have some form of venting system to allow air exchange and prevent the build-up of harmful gasses. A vent pipe should extend a minimum of 6 inches (15 cm) above ground from the top of a buried tank. A standard vent size is

STEP 2: DESIGN

3 inches (7.5 cm) and should have a barrier to prevent the entry of small animals or insects. A gooseneck fitting with a screen is the most straightforward way.
- Buried tanks need a properly excavated hole, tank bedding for support, and proper backfilling. These requirements will be based on the manufacturer's specifications and must meet local codes and standards for buried storage tanks.

Connecting to the Transfer System

The Transfer System should enter the Storage Tank at least 2 inches (50 mm) above the Overflow drainage piping. This connection should be based on the manufacturers' specifications for the Storage Tank.

Cold Weather Modifications

If you are not burying your Storage Tank below the frost penetration depth, here are your options for cold weather modifications to prevent freezing.

Disconnecting during Winter

This option diverts water away from the Storage Tank during the winter months. Then, the Transfer System is either disconnected entirely (returning to your original gutters and downspouts), or a bypass is added to reroute water away from the Storage Tank. Disconnection also requires a drain valve on the bottom of the Storage Tank so it can be emptied.

This is the cheapest and easiest option if your usage is mostly in summer and you don't need water to be collected in the winter.

Winterizing

There are a few ways to keep your Storage Tank temperature above 32°F (0°C). Remember that all outdoor RHS components, not just the Storage Tank, may need cold weather modifications.

- Insulation: if you only have freezing temperatures intermittently, you might be able to get away with adding insulation around your Storage Tank. While styrofoam insulation is a standard option, you could also build a small shed or structure around the tank to isolate it from cold weather, heating that space as necessary.
- Heating: adding a heating system inside your Storage Tank will require an additional electrical connection and ongoing maintenance costs. But, this is the most reliable way to ensure it does not freeze.

PUMP

Pumps can fit inside the Storage Tank or may be placed nearby and supplied water via gravity from the Storage Tank. Pumps will require supporting Sensors and Controls and an

STEP 2: DESIGN

electricity supply. Pumps are not mandatory, and Storage Tanks can be used with a gravity-fed faucet, spigot, or even a hand pump for a more straightforward approach. However, a pump is recommended if you need pressurized water or if your desired point-of-use is further away from your Storage Tank location.

The Pump System has two main parts: the pump itself and a pressure tank. The pump creates the pressure, and then the pressure tank stores the pressurized water; this allows constant water pressure to be maintained within the Connection System.

Pump Systems can also be purchased as all-in-one packages, with components guaranteed to work together and intended for use in an RHS. If you are not feeling confident in selecting the individual components of a Pump System, consider purchasing a Pump System package specifically intended for this purpose.

The Pump System needs to supply water to all connected fixtures at a sufficient rate and pressure to meet those needs. If you have a single connection, this is reasonably straightforward to figure out; however, if you plan to connect multiple fixtures, this can become very complex very quickly.

Homeowner maintenance is especially critical for pump systems; you should be capable of basic troubleshooting if there is an issue with the pump.

Types of Pumps

There are three type styles of pump to choose from:

- Constant speed - once the pressure drops in the Connection System, the pump will turn on until a set pressure level is reached. These pumps are cheaper and suitable for an RHS without large pressure differences (e.g., garden hoses and refilling toilet tanks) but can require larger pressure tanks.
- Variable speed - these pumps decrease or increase their speed to maintain a consistent level of water pressure. They are more expensive but smaller, require less power, and deliver constant pressure.
- Manual - are less convenient but much less expensive and easier to maintain. A simple hand pump solution may be best if you use water for landscaping or gardening.

Small water pump

STEP 2: DESIGN

Important: For large-scale applications, secondary pumps (or jet pumps) may be needed to repressurize the Connection System or to deal with elevation changes. For example, if you are supplying water throughout a multi-story building. Typically these solutions are not necessary for residential applications, so they are not covered in this book.

The last factor is the potential for downtime. If occasional interruptions in pressure are acceptable (for example, flushing an outdoor toilet), then a standard pump system will work fine. However, if interruptions are unacceptable (e.g., providing drinking water), then a dual-pump or automatic bypass system are options. Unfortunately, these options are both expensive. A dual pump system requires a backup pump to be built into the Connection System. A bypass system uses a separate water connection so that an alternative water source can be accessed if necessary. Neither of these options is recommended unless uninterrupted service is essential. Still, a dual pump system is a much simpler option between the two.

Pump Requirements

Once you know what type of pump you need, the next step is to calculate the required flow rate.

The easiest way to select the flow rate is to sum the required flow rates of all fixtures attached to your RWH system. A worst-case scenario assumes all fixtures are open and running simultaneously. This calculation is the easiest but will greatly overestimate your flow needs. A better approach for complex Connection Systems is to calculate the maximum peak flow and buy a pump to handle a percentage of that. 80% is a conservative estimate that will ensure you have pressurized water almost all the time.

The following variable in selecting a pump is the pressure or head, and two factors must be considered:

1. required system pressure – This is the pressure required by the fixtures connected to the pump and pressurized Connection System; and

2. total dynamic head – This is the loss in pressure (or "head loss") that occurs when water is lifted from a low elevation to a high elevation, and the loss that occurs when water is being pumped through a long stretch of pipes and fittings.

Total dynamic head is composed of three components:

1. Static lift is the height the pump must lift water before arriving at the pump. This factor is only applicable to systems utilizing a secondary pump, so it is not relevant to us

2. Static height is the height from the pump to the furthest vertical fixture

3. Friction loss is the pressure loss when water travels through pipes and fittings

STEP 2: DESIGN

Important: Calculating the required pump flow and pressure can be challenging. It is strongly recommended that you confirm your intended use with the manufacturer to see if a pump will meet your needs. Most pumps will specify the flow rate they can supply and provide a head or vertical lift value. Compare the flow rate the pump can deliver to your fixture (or to 80% of the total flow rate for multiple fixtures). And as a general rule of thumb, you need at least 4 inches (10 cm) of vertical lift for every 3 feet (1 m) of horizontal pipe in your Connection System.

Generally, most outdoor watering and indoor uses require a minimum of 25 PSI pressure; however, some uses, like drip irrigation, may need as little as 10 PSI.

Pressure Tanks

Select a pressure tank compatible with the pump type and the pump's flow rate. Ideally, you would select a pump with a built-in or recommended pressure tank from the same manufacturer to ensure they work together.

Constant speed pumps require larger pressure tanks designed to store water discharged by the pump over a 1 to 2-minute period, based on the pump's minimum run time. A longer pump run time requires a larger pressure tank but minimizes wear due to less frequent pumping.

Variable speed pumps can use pressure tanks as small as 1 USG (4 liters), but the pump manufacturer will typically specify the requirements.

Selecting your Pump

Follow these steps to choose an appropriate pump:

- Figure out fixtures to be connected and their required flow rates and pressures
- Choose between constant and variable speed
- Determine the required flow rate and pump head
- Consult applicable codes and regulations for applicable requirements for minimum or maximum pressure
- Consult the pump manufacturer or supplier, or use the pump manufacturer's 'pump curve' charts to select the appropriate pump model
- Select a compatible pressure tank based on the pump manufacturer's specifications
- Plan layout for the pump, pressure tank, and distribution system

Important: Dry-running is when the pump is running with no water and can cause pumps to get overheated and damaged. The pump intake within the Storage Tank should be below the tank's low-water level and below the point where a sensor activates a Make-up system or automatic

STEP 2: DESIGN

pump shutoff. Selecting a pump with dry-run protection is an option if your RHS has no sensors or controls to prevent dry-running.

OVERFLOW

If your Storage Tank overflows, you need to control where all the extra water ends up. Without an overflow system, that excess water could back into your Transfer System, flood the area around your Storage Tank, or cause other damage to the RHS or surrounding structures.

While an overflow system is a requirement of RHS, there are various levels of complexity. At the simplest, an Overflow system could be an additional pipe or hose that drains excess water away from the Storage Tank. However, local regulations and codes may require more active overflow management, including treatment and best practices to mitigate the environmental impacts of stormwater runoff.

There are five options to direct overflow to a suitable location:

1. Discharge to grade via gravity flow: this works for above-ground and buried tanks near a hill. Overflow drainage piping directs water above the high water level in the Storage Tank outside to a well-drained ground area, away from buildings or areas at risk of

erosion. This option is the most straightforward and recommended.

2. Discharge to grade via pump-assisted flow: this works for buried tanks, where overflow has to be pumped back up to the surface. A pump will be located within the Storage Tank, activated once a sensor determines water has reached the high water mark.

3. Discharge to a storm sewer via gravity flow: this works for above-ground and buried tanks, but you will need to check your local regulations for discharging into storm drains. In most cases, Storage Tanks cannot be directly connected to sewers and require an air gap or other indirect connection. If there are rainwater gardens or other forms of stormwater management in your area, you will need to confirm if you can use these areas for overflow.

4. Discharge to a storm sewer via pump-assisted flow: this works for above-ground and buried tanks. In situations where the storm sewer drain is above the tank, you will need to check your local regulations for discharging into storm drains.

5. Discharge to soakaway pit via gravity flow: A soakaway pit is an excavated area filled with sand at the bottom, then crushed stone or gravel with the overflow pipe directed into it, then covered in soil to the surface surrounded by filter fabric. The crushed

stone or gravel gives the pit structure and slows the passage of water downwards. The sand layer at the bottom filters the rainwater before leaving the pit. The overflow pipe in this section will have holes to distribute water evenly throughout the pit. The outer fabric layer prevents soil outside the pit from entering and clogging up the gravel/stone and sand layers. Soakaway pit design is highly dependent on your local soil conditions. They require considerable space and an understanding of your soil permeability. Depending on where you live and amount of space you have, soakaway pits may not be a feasible option.

Relying on pumps to manage overflow is generally not recommended, as this is an additional point of failure in the RHS.

While *discharge to grade via gravity flow* is recommended, there are a few things to check first:

- Are there any applicable local water management requirements? Retention time, treatment requirements, and infiltration or overland flow could all be requirements.
- What is your terrain like? If you are in a flat or depressed area, or if your soils are heavy clay and don't drain well, there may be nowhere for this water to go. Overflow needs to be discharged to areas with

no risk of pooling. To prevent erosion, you should also avoid discharging overflow to bare soils and steep slopes.
- Are there buildings nearby? Do not direct overflow towards any existing structures.

If discharge to grade isn't an option, you may need to discharge directly to a storm sewer. In that case, there may be additional restrictions or regulations depending on where you live.

Size

Your overflow pipe needs to be equal to, or larger than, the Transfer System piping.

Placement

The Overflow piping should exit the Storage Tank approximately 2 inches (50 mm) below the bottom of where the Transfer System Pipes enter the Storage Tank. This placement ensures that the Transfer System will not get backed up during heavy precipitation events.

The Overflow pipes should discharge into an area with low erosion risk. Low-risk areas could include spots:

- with heavy vegetation
- covered with a concrete splash pad
- with piles of small rocks and stones

STEP 2: DESIGN

Simple Overflow system

MAKE-UP SYSTEM

There may be times when your Storage Tank runs dry. Lack of water can be a problem if you rely on your RHS as your primary water source, and dry conditions can also damage the Pump System. If either of these applies to your RHS, there needs to be a trigger for either a warning system (like lights or automated message) or a way to switch to an alternative source automatically. A Make-up system is optional if you are not using a Pump and have other water sources.

Make-up systems allow water from other sources to be added directly to the Storage Tank. However, they require water of at least the same quality or better than rainwater, typically from domestic water sources or municipal connections.

Type

There are two types of Make-up water systems:

Top-up - another water source is used to fill the Storage Tank. These can be simple, like manually filling up the Storage Tank yourself, or complex, with direct connections to other water sources. Direct connections to other sources usually require an air gap to be in place.

Bypass - the Pump system within the Storage Tank is shut off, and another source supplies the connected fixtures. This option is usually only necessary if you provide indoor fixtures or rely on a regular supply of rainwater for automated applications, like a timed irrigation or watering system.

Top-up style systems are allowed more often in plumbing codes since bypass systems might contravene rules against a non-potable source being connected to a potable one. Therefore, this section will focus only on top-up systems. These can be either:

- Manual top-up - requires you to put water into the Storage Tank directly. There may be interruptions if the Storage Tank runs dry, but it only requires basic monitoring to put this in place, like regularly checking tank levels. Recommended if you are also manually controlling the use of your rainwater.
- Automatic top-up is more complex and usually requires power, a control system, and an air-gapped connection to another source. Recommended if another automated system also manages your

STEP 2: DESIGN

rainwater use. This option would provide dry run protection for your Pump, as the pump intake is located below the low water level, activating the Make-up system.

The following control equipment is needed for an automatic top-up system:

- water level sensors located in the Storage Tank
- solenoid valve located on the Make-up water supply pipe
- A minimum 1 inch (25 mm) air gap between the supply pipe and high water level
- piping conveying make-up water to the Storage Tank
- electrical conduits containing wiring from water level sensors and pumps.

A float switch will turn on when the switch is down and activate when water in the Storage Tank reaches the low water mark. Once the float switch is activated, it will supply power to the solenoid valve, which is normally closed and will open the valve. The open valve will allow water from the secondary source to flow into the Storage Tanks.

This water will raise the float switch back up and automatically remove power from the solenoid valve, returning it to the closed position. The volume of make-up water added is based on the length of the float switch tether. The tether length should be as short as possible to maximize rainwater use. The tether point should be as low as possible while providing just enough water for the Pump. For most Storage Tanks, a recommended tether length of 3 inches (75 mm) works well.

Size

Make-up pipes should be large enough to handle the total flow rate from the alternate source. Make-up systems typically use gravity flow to prevent backup, so every element needs to be sized the same.

Avoiding backflow

You can use an air gap between the make-up and rainwater harvesting systems to avoid backflow. Air gaps must be visible, unobstructed, located above the Overflow, and a minimum of 1 inch (25mm) or twice the diameter of the water supply pipe.

TREATMENT SYSTEM

A universal treatment consideration is the maximum allowable amount of sediment in the treated water to prevent clogs or build-up within the connected fixtures.

If you are planning to use rainwater as drinking water or for any domestic purposes, you will need a Treatment System that is rated to meet your applicable drinking water guidelines. These guidelines may also be called drinking water objectives or regulations, and this water can also be called potable or domestic.

Treatment requirements should be determined on a case-by-case basis by local building or health authorities, considering the designers' recommendations and end users' preferences.

Treating rainwater after it leaves the Storage Tank is technically optional; however, as a general rule, only the following fixtures should be connected to an RHS without treatment:

- toilets or urinals
- directly connected underground irrigation systems

Check your local plumbing and building codes to confirm allowable uses for untreated rainwater.

Treatments that address chemical and biological contamination are more complex and may require additional power, filters, chemicals, or other materials. Treatment Systems may also require pumps and electrical mechanisms to facilitate treatment, unlike the Pre-filter, which can rely only on gravity flow.

Similar to Pump Systems, Treatment Systems are often sold as an all-in-one package specifically for use in an RHS. This option is recommended, as Treatment Systems can be very complex.

Designing your own Treatment System and selecting the individual components while still maintaining the desired level of water quality is a highly complex process. Instead, find a Treatment System package that will meet your needs based on the information provided below.

Types of Contamination

Physical - dirt, debris, and dust will make drinking your water cloudy and unpleasant. If you are not drinking your rainwater, sediment can clog up fixtures and damage pumps or other RHS components. Physical contamination, like sediment or solid particles, can also carry biological and chemical contamination. Therefore, removing physical contamination is crucial to keeping your RHS and your family safe.

Physical contaminants will always be present to some degree. However, you may experience more if you receive infrequent rains, live in a dusty area, or are near a major roadway or industrial location.

Biological - this includes bacteria, viruses, or parasites (pathogens) that may have washed off your Collection Area into your Storage Tank or may be growing in an improperly sealed RHS. There are not always obvious signs, and these pathogens usually only result in mild illness if ingested, but in the worst case can be fatal.

Pathogens can survive on plants, so if you are using your rainwater in a garden, removing as many pathogens as possible is

STEP 2: DESIGN

recommended. If removing pathogens is not possible, please take extra care to clean any harvested food properly before eating.

The primary source of biological contamination is wildlife defecating onto your roof or finding ways to access your Transfer System.

Chemical - While chemical contamination is a big concern in drinking water, there is limited opportunity for the chemical contamination of rainwater. Chemicals are not a priority as long as your roof surface is clean and you use recommended materials for all RHS components. If you plan on drinking harvested rainwater, make sure you get proper laboratory testing done to ensure it is safe and does not require further treatment. Request analysis for the following contaminants from your local environmental laboratory: routine chemistry, metals, and polycyclic aromatic hydrocarbons.

The primary source of chemical contamination is physical contamination washed into your Transfer System and Storage Tank.

Types of Treatment

Filtration - this removes anything suspended in water, including sediment that can clog fixtures or carry other contaminants. Filtration deals with physical contamination, but some types of filtration can also remove chemicals or pathogens. A basic filtration system can drastically increase the quality of your harvested rainwater.

- Mechanical filtration - a fine mesh that prevents particles from crossing.
- Sand filtration - passing water through a layer of fine sand can also remove some chemicals and pathogens.
- Membrane filtration - a permeable membrane that water can pass through but not much else; this can also remove pathogens.
- Reverse osmosis - forces water through a very fine membrane at high pressure and can remove metals and smaller pathogens. Unfortunately, this method results in significant loss of rainwater and is costly. However, reverse osmosis systems can be installed in small spaces, like directly under a kitchen or bathroom sink.
- Activated carbon - are filters that trap organic contamination and can also remove bacteria.

Filtration to at least five microns is recommended. Anything larger than this size can harbor pathogens, and if larger particles are still present in water, they can prevent proper disinfection and increase wear on fixtures.

It is possible to incorporate filtration into your RHS before the Storage Tank. Gravity-fed filters, such as *slow sand* filters,

STEP 2: DESIGN

are inexpensive, require no power, and could be a DIY project. These filters should contain fine sand, ideally sold as *filter sand*, but any sand will work as long as it has been thoroughly rinsed and cleaned. The depth of the sand in the filter will determine how much it cleans the water, and the diameter will determine the flow rate (slower is better). A small 5-gallon filter will remove some contamination, and a larger 55-gallon drum will be able to remove most particles and even some pathogens.

While these are cheap treatment systems, and you could even build one yourself, they are extremely heavy and difficult to move once in place. Unless you are intended to generate potable water, a first-flush system is a much simpler solution.

Disinfection - can be used to remove biological contamination, including bacteria, viruses, or other parasites. Unless you are drinking this water or watering edible plants, this is not necessary. Proper disinfection requires that the water has been adequately filtered first, and no disinfection method works well if the water is cloudy or contains too much sediment. Common examples of disinfection methods are:

- UV light - direct exposure to intense UV light can kill most pathogens, depending on how long the water is exposed to the light. UV disinfection systems can be added directly to pipes and require minimal power.

- Chemical - adding chemicals like chlorine or ozone to rainwater can disinfect it but requires regular addition of chemicals and rigorous testing. This option is not recommended unless you have professional experience with water treatment.
- Thermal treatment - the easiest way to remove pathogens is to boil the water before using it. Caution is needed for this approach, as under certain conditions, this can result in you increasing the concentrations of other chemical contaminants.

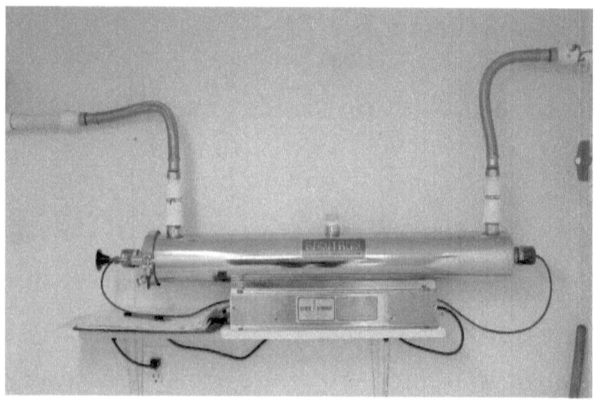

Small-scale UV light disinfection system

A second option for disinfection is to remove these pathogens *at the tap*. Often called *point-of-use* systems, this may be a more practical option if there is only one fixture supplying water used for drinking or cooking. Alternatively, boiling could be used as a final disinfection step if the water has been appropriately filtered.

STEP 2: DESIGN

Treatment for Potable Use

If you intend to treat rainwater for use as potable water, it is strongly recommended to send a sample of collected rainwater to an environmental laboratory for analysis first. These results will determine what contaminants may be present and allow you to select an effective Treatment System.

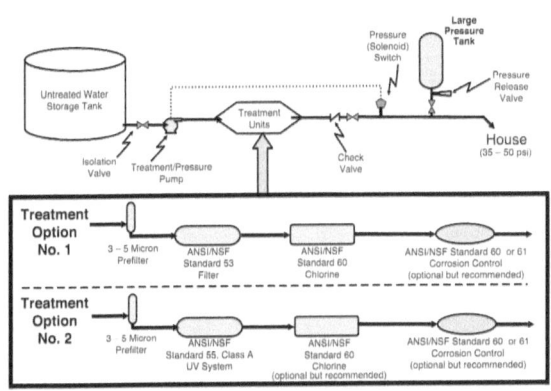

Example diagram of a home drinking water treatment system

Treatment for Non-Potable Uses

Treatment requirements are less important for non-potable uses. Still, some form of filtration is always recommended to remove pathogens and larger debris that can block fixtures or build up in other parts of the RHS.

Treatment System Size

Treatment devices should be sized following the maximum flow rate of the Pump System and the manufacturer's requirements.

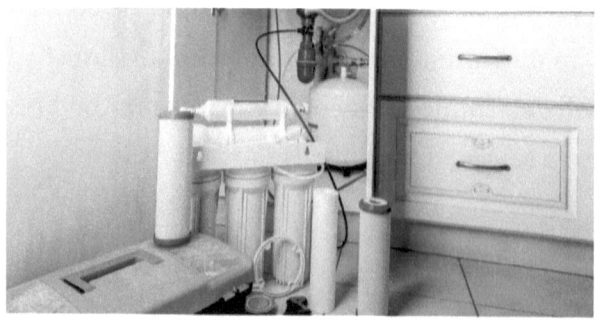

All-in-one treatment system to install underneath a sink

CONNECTION SYSTEM AND FIXTURES

Connections between the Storage Tank and fixtures are made with pipes. Gravity-fed Connection Systems are simple and can be designed the same way and with the same requirements as the Transfer System.

Pressurized Connection Systems are more complex, and have two distinct parts. First is the service pipe, which extends from the Storage Tank to the pressure tank. The second is the supply pipe, which continues from the pressure tank to the fixtures.

These pipe sections will require different materials, sizing, and installation requirements.

STEP 2: DESIGN

For example, service pipes will always have water inside and are not pressurized, usually with a larger diameter than supply pipes. However, the supply pipes are pressurized, may be within buildings, and will require protections against cross-connections if there are multiple fixtures.

Service Pipes

Service pipes should be horizontal or on a slight upward slope to the pump (0.5-2%), so they drain back into the Storage Tank if the pump is not active. The best practice for service pipes is to use as few pipe fittings as possible between the Storage Tank and the pump.

Supply Pipes

You should select supply pipes based on pressure requirements from your Pump System.

If you are planning to have multiple Fixtures or a more complex Connection System, asking for advice from an irrigation or landscaping contractor, your pump supplier, or a professional plumber is a good idea.

Fixtures

Taps, faucets, spigots, toilets, and appliances are all potential fixtures that could use rainwater. The fixture at the end of the Connection System needs to be appropriate for the water's quality, flow, and pressure.

Taps in kitchens or fixtures for human consumption require water treated to drinking water standards.

Most fixtures require a minimum flow and pressure, which you should follow to ensure proper operation. Generally, a flow rate of 4 gallons per minute (15 L/min) and pressure above 35 PSI will work for most indoor and outdoor fixtures.

For any fixture receiving non-potable water, it is strongly recommended to include a clear label indicating that the water is unsafe to drink.

SENSORS & CONTROLS

The types of sensors and control needed will be based on what automatic functions your RHS needs. Ensure any sensors and controls are supplied with sufficient power to operate and control mechanical parts, such as valves.

The most commonly used sensor is a float switch. This type of switch connects to a floating tether, turning the switch on or off depending on the water level. For example, float switches can be connected to pumps for automatic shutoff at low water levels to prevent dry-running or automatic operation at high levels as part of an overflow system.

Any sensors or controls intended for use in the Storage Tank should be able to work while submerged and should be installed within easy reach of the Storage Tank opening.

STEP 3: SELECTING COMPONENTS AND SAFETY CONCERNS

I t is best if you have a solid idea of what components your RHS needs and the types of features you would like to have before making major purchases. The following section will help you select specific parts, ensure that these are all compatible with each other, and highlight any safety or engineering concerns you should be aware of before making your final choices.

Depending on the availability of parts and current costs, you may have to adjust your designs again during this step. The prices and availability of these materials can vary considerably, so you will have to spend some time and effort sourcing reliable suppliers.

Purchasing an all-in-one system or hiring a contractor to select and install the parts for you is an option; however, I would still strongly encourage you to read through this step so that you can ask the right questions and be an informed consumer.

While it is not always possible to do so, developing a purchasing plan that includes all RHS components before buying is strongly recommended. This plan will allow you to ensure that all parts are compatible and of sufficient quality to meet the goals you have for your RHS, and if necessary, adjust your design and budget accordingly.

SELECTING YOUR RHS COMPONENTS

As a general rule, ensure that all your RHS components are made of chemically inert and non-toxic materials.

Common styles of approved certification logos

Collection Area (roofing material)

Replacing your roof material is a massive undertaking in terms of time, effort, and financial cost. There are many considerations in selecting an appropriate roof, and compatibility with an RHS is only a part of the overall decision. If you

STEP 3: SELECTING COMPONENTS AND SAFETY CONCERNS

are already planning on replacing your roof and are looking to incorporate rainwater harvesting, ensure that the new roof material meets the following conditions:

For harvesting potable water: NSF/ANSI 61

For harvesting non-potable water: NSF P151

Most gutters and downspouts on existing roof areas are made from aluminum or galvanized steel, which are considered suitable for an RHS. Copper, wood, vinyl, and plastic materials are not recommended.

> *Important*: Green roofs, covered with vegetation, are great for the environment but not suitable for collecting potable rainwater. These roof types can introduce organic compounds into rainwater. While these compounds are relatively harmless by themselves, they can create dangerous byproducts when mixed with chemicals used for disinfection treatments.

If you are installing new gutters, ensure that all gutters are supported by hangers or other physical restraints spaced approximately 18 inches (450 mm) apart.

Pre-filter

Gutter guards, leaf filters, and other basic Pre-filters are available from several retailers and can be sourced and installed yourself or by gutter contractors.

Pre-filter devices must be sized to handle the peak runoff from the Collection Area based on the likely worst-case precipitation event expected in your area.

A combination of one coarse and one fine pre-filter is recommended. The coarse filter will remove larger objects like leaves, twigs, and small stones, leaving the fine filter to more easily remove dust and smaller contaminants. This setup will also reduce the maintenance necessary for the Pre-filter system, as fine filters or screens require more regular removal of blockages.

Select a metal (typically aluminum) version with a size that is compatible with your gutter type. Pre-filters should primarily be chosen based on the debris you are most concerned about: leaves, dust, or animal waste.

Transfer System

Several criteria must be considered when selecting pipe material. First, the chosen pipe must be rated as suitable for *ultraviolet* (UV) light exposure if above ground or *burial* if below ground. Ensure that the burial depth the pipe is rated for is consistent with your planned burial depth. Using ABS or PVC sewer piping above ground is not recommended unless it has been painted to protect it from sunlight.

If you are using the collected water for domestic purposes, high-quality materials meeting NSF P151 certification is recommended. In general, a type of polyvinyl chloride (PVC)

STEP 3: SELECTING COMPONENTS AND SAFETY CONCERNS

pipe, referred to as "sewer grade pipe" or "PVC SDR35" is recommended for all RWH systems, as it meets these criteria.

Acrylonitrile-butadiene-styrene (ABS) is another type of pipe. It is typically less expensive than PVC SDR35 but may not be appropriate for all RWH systems as it is not rated for any level of UV exposure.

PVC DWV pipe is a good choice for above-ground applications. However, potable water-rated PVC Schedule 40 piping is also recommended if you use rainwater as drinking water.

PVC sewer-grade pipe is recommended for below-ground applications. ABS sewer pipe is a cheaper option but should only be used if joints between pipe lengths are cemented to avoid leaks and cannot be used above ground. In areas where heavy weights may be applied (e.g., under driveways) or pipes are buried to shallow depths (less than 2 ft or 60 cm deep), PVC SDR or Solid Core ABS should be used.

The selected pipe must also be approved for use in an RHS by the applicable local codes and regulations. Therefore, purchasing pipes specifically designed for an RHS is strongly recommended.

Cleanouts are required on Transfer System pipes to allow for regular cleaning and removal of any debris or clogs in the system. Check your local building or plumbing codes for cleanout spacing and size requirements. A general rule for spacing cleanouts is every 50 feet (15 m) *and* every time you change direction.

Storage Tank

A Storage Tank that meets NSF/ANSI Standard 61 certification is recommended if you are using rainwater inside your home. Otherwise, select an RHS designed for storing rainwater or potable water use. Storage Tanks made with plastic, food-grade plastic, reclaimed plastic, or steel are standard options.

Plastic, usually polyethylene or polypropylene, rain barrels are the most common. Lightweight, low maintenance, and inexpensive, they are an excellent choice for a simple and compact RHS. Ensure they are food-grade and UV treated.

Fiberglass tanks are usually only found in commercial or industrial applications. They require an additional interior food-grade liner if being used for potable water.

Precast concrete tanks are extremely heavy and have a long lifespan. Therefore, they are only recommended for residential applications if they need to withstand high pressure (e.g., placed below ground under driveways).

Poured concrete tanks are excellent for custom solutions and complex spaces but will require additional design support from a qualified professional.

STEP 3: SELECTING COMPONENTS AND SAFETY CONCERNS

Steel tanks are an excellent option for large above-ground tanks, with many size and shape options available. In addition, this style of tank may be able to be assembled onsite for difficult-to-access areas.

Barrels can also be made of wood (oak, beechwood, cedar, or hickory) to add a more decorative touch. These materials are not suitable for potable water use but are fine for landscaping and gardening. However, wood barrels are heavier, more expensive, and require regular maintenance to check for deterioration, algae growth, and rot if they do not have a protective coating.

Storage Tanks should have an access opening to install, inspect, and maintain any internal components (pumps, heaters, etc.). Confirm sizing with your local codes and standards, but these openings should be at least 18 inches (45 cm) wide and have drip-proof, non-corrosive covers. A tank with an opening larger than 4 inches (10 cm) should have a lockable cover.

Storage Tanks can come in different shape configurations, so ensure that the volume of tank you are looking for also has a footprint compatible with the space you have available.

Also, make sure the Storage Tank you choose has all the inlets and outlets you need. At a minimum, you need an inlet for the Transfer System and outlets for the Overflow and

Connection Systems (these can be cut into a tank if necessary).

The placement of inlets and outlets will also influence the Storage Tank volume. All space above the Overflow and below the outlet to the Connection System should not be included when determining the *operational* Storage Tank volume. The actual capacity of a Storage Tank is often between 85% and 95% of the total Storage Tank volume.

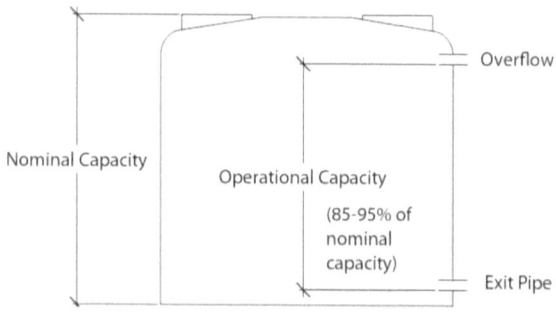

Operational capacity of a Storage Tank

Important: Ensure you are using the correct units when selecting a Storage Tank and not mixing Imperial gallons with US gallons.

Overflow and Make-up Systems

Piping made from PVC SDR35 or ABS is recommended for gravity flow systems and polyethylene pipes for pumped systems. Pipes used in the Overflow and Make-up Systems should be as big or larger than Transfer System pipes.

STEP 3: SELECTING COMPONENTS AND SAFETY CONCERNS

Automated Make-up systems use a *normally closed* (NC) float switch with an NC solenoid valve so that the Make-up system is not operating when water is above the low-water mark.

Pump

A basic all-in-one Pump System could cost in the range of $500 to $2,000.

Ensure that the Pump System pressure or 'head' is sufficient to move the water through the Connection System at the desired flow rate and pressure.

Also, check that the power supply supports all selected RHS components, including the Pump System, sensors, and controls.

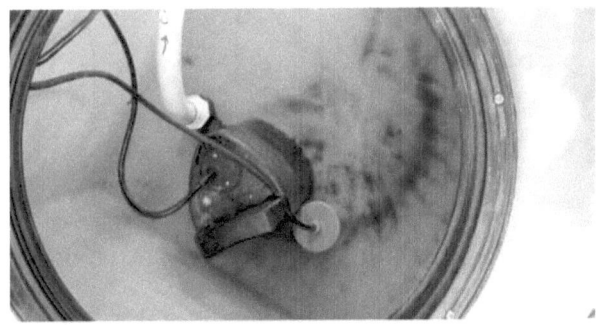

Simple Pump System in a small Storage Tank

Connection System

Cross-linked polyethylene (PEX) pipes are recommended for Connection Systems that use pumps, as this material is better able to handle pressurized water. Ensure that any piping is rated to take the pressure supplied by the Pump System and pressure tank.

Cleanouts are also required on Connection System pipes to allow for regular cleaning and removal of any debris or clogs in the system. Check your local building or plumbing codes for cleanout spacing and size requirements.

Important: If you are only interested in collecting rainwater in a basic barrel system, many Storage Tanks designed for home use come with a standard garden hose connection built in as an outlet near the bottom. So, no additional Connection Systems or Fixtures are necessary.

Treatment System

This system must be able to hook up to the Connection System and be compatible with the pressure and flow of the Pump System. Purchasing an all-in-one package designed explicitly for rainwater is strongly recommended. The Treatment System should indicate what water quality standards it can meet and have certification to your country's drinking water standards.

STEP 3: SELECTING COMPONENTS AND SAFETY CONCERNS

Some Treatment Systems are intended to be placed near the Pump, while others can be placed inside the home just before the fixture that needs potable water (the point-of-use).

Take note of any requirements for input water quality, as additional filtration or other treatment, may be necessary to avoid damaging the Treatment System.

Ensure that any treatment device has the proper certification. For example, for potable water use, look for NSF/CSA 61, drinking water filtration NSF/ANSI 42 & 53, and UV treatment NSF/ANSI 55.

If you are going to use a filtration system, here are some common filter sizes for reference, along with the size of contamination they can remove:

- US Mesh #18 = 1,000 microns - approximately the eye of a needle, small stones
- US Mesh #25 = 707 microns - beach sand
- US Mesh #35 = 500 microns - human hair or sawdust, construction dust
- US Mesh #45 = 354 microns - coffee
- US Mesh #60 = 250 microns - fine sand
- US Mesh #80 = 177 microns - dust mites
- US Mesh #400 = 37 microns - pollen or dust
- 5 microns - required for UV light disinfection
- 1 micron - bacterial or parasitic pathogens

Sensors and Controls

Ensure that the power supply supports all selected RHS components, including the Pump System, sensors, and controls. Also, check that it can supply water to wet or submerged equipment.

Fixtures

Fixtures can operate on rainwater, and requirements for filtering or sediment content are compatible with the RHS filtrations method.

Check fixtures' pressure and flow requirements against what your RHS can supply, or select the Pump System based on fixture requirements.

COMPATIBILITY CHECKLIST

Collection Area and Pre-Filter

Any Pre-filters are compatible with your existing gutter system materials and size.

Collection Area and Transfer System

Transfer System pipes can connect with your existing gutters and downspouts, or you have found an adaptor that will allow this connection.

STEP 3: SELECTING COMPONENTS AND SAFETY CONCERNS

The Transfer System pipe size is consistent with the maximum expected rainfall.

Transfer System and Pre-FIlter

Any Pre-filters are compatible with your planned Transfer System pipes.

Transfer System and Storage Tank

Transfer System pipes can connect with appropriate Storage Tank openings, or adaptors can secure this connection.

Storage Tank and Sensors & Controls

Necessary sensors and controls can fit inside the Storage Tank and are accessible for inspection and maintenance.

Required electrical connections for Sensors and Controls can be routed into the Storage Tank as necessary.

Storage Tank and Connection System

Connection System pipes can connect with appropriate Storage Tank openings, or there are adaptors that can secure this connection.

Storage Tank and Pump System

The Pump System can fit inside the Storage Tank, or the Pump System can operate based on the Storage Tank's natural water flow.

Pump System and Treatment System

Water pressure generated by the Pump System is compatible with Treatment System requirements.

Pump System and Fixtures

The pressure produced by the Pump System will be sufficient for desired Fixtures.

Pump System and Connection System

Connection System pipes can handle the pressure produced by the Pump System and have a compatible size.

Connection System and Fixtures

Connection System pipes can attach to the selected fixtures.

Treatment System and Fixtures

The Treatment System produces water of sufficient quality for water use of Fixtures.

SAFETY CONCERNS BY RHS COMPONENT

Collection Area

Is your roof material suitable for capturing safe drinking water?

Do you have the proper equipment to safely access and clean your roof when necessary?

STEP 3: SELECTING COMPONENTS AND SAFETY CONCERNS

Pre-Filter

Do you have the proper equipment to safely access Pre-filter components for installation and maintenance?

Will your Pre-filter result in clogs or back up your Collection Area gutters or Transfer System that could damage your home?

Transfer System

Are all elevated portions of your Transfer System supported and able to withstand extreme weather?

Are any below-ground pipes properly backfilled and marked?

If you are cutting into a metal downspout, wear heavy gloves to avoid cuts and safety glasses to prevent eye injuries.

Storage Tank

Are there safety measures to prevent small children from entering the Storage Tank?

Are there warning labels displaying *DANGER-CONFINED SPACE* at the maintenance opening?

Aboveground Storage Tanks must be covered and tightly sealed to keep out pests like mosquitos and also prevent your tanks from becoming a drowning hazard for small children or pets. In addition, if sunlight can reach water in the Storage Tank, this can promote algae growth within your RHS.

Is all interior maintenance done by individuals with enclosed space training (where applicable)?

Is the Storage Tank placed on secure, stable ground?

If below ground, is the Storage Tank properly backfilled and marked?

Does the manufacturer intend it for use as a rainwater collection Storage Tank?

> *Important*: It may be considered' confined space' work if you need to install components within the Storage Tank. Entering a Storage Tank yourself is not recommended and should only be undertaken following local confined space entry regulations.

Wherever possible, you should install interior components within easy reach of an access port. If this isn't possible, they should be installed by the manufacturer or someone trained in confined space work.

Pump, Sensors, and Controls

Are all electrical connections waterproof and regularly maintained?

If there is an external pump, does it have its dedicated electrical circuit and a dry, well-ventilated location?

STEP 3: SELECTING COMPONENTS AND SAFETY CONCERNS

If there is a submerged pump, does it have a ground fault-protected electrical circuit?

Does the Pump System have dry-run protection or a sensor to prevent operation at low-water levels?

Overflow

Is overflow water directed away from building foundations and other structures without creating pools or standing water?

Treatment System

Is water quality tested regularly to ensure it is appropriate for the desired end uses?

Are disinfection chemicals applied by qualified persons and tested regularly?

Connection System

Are connections completely sealed and watertight?

Are all aboveground elements of the Connection System supported and able to withstand extreme weather?

Fixtures

Do not drink or give untreated rainwater to pets. If you use rainwater for edible garden crops, use drip irrigation or soaker hoses and apply the water directly to the soil and not onto the plants themselves.

Important: Always wash fruits, herbs, and vegetables with clean potable water before you eat them.

STEP 4: INSTALLATION AND TESTING

Most of the complex components of an RHS can either be installed by a qualified professional or will include detailed instructions from the manufacturer. The most important part of the installation is developing a plan and logical installation order.

The Storage Tank is the central hub of your RHS, and installing that first will make things easier. After your main tank is installed, any secondary tanks, like pressure tanks or filtration treatment systems, can be placed accordingly. Then you can build in the Transfer and Connection Systems on paths that have been finalized. It is much easier to adjust the placement of pipes than to adjust the placement of a large Storage Tank.

FINDING PROFESSIONALS

Depending on the complexity of the RHS you designed, you may want to get professional help or advice from local experts who know the relevant building codes and regulations like the back of their hand. Finding local experts is vital, as they will understand the climate, soil conditions and may also know what has and hasn't worked for others in the past.

Here are some places you can start looking:

- Municipal building inspectors: if you want to know what building codes and regulations might apply to your RHS
- Builders: might be willing to give someone advice or recommend reputable contractors, or ask about any engineering or structural concerns
- Product suppliers: asking the manufacturer about your design and if they foresee any problems is a good idea, and if they are aware of any issues using their products where you live
- Plumbers: an RHS is mostly plumbing, so any advice you get from a professional regarding RHS design, material selection, or potential issues should be taken seriously
- Landscaping professionals: may have experience working with an RHS or a better idea of realistic water requirements for non-potable outdoor use.

STEP 4: INSTALLATION AND TESTING

- Drinking Water Officer: if you are planning to treat rainwater so that it is safe to use as drinking water, check with your local environmental, health, or drinking water branch of your local government. They will let you know what requirements you have to meet and may be able to suggest treatment systems or review your design to ensure it meets the legal requirements.

INSTALLATION CHECKLIST

As a general rule, all components of your RHS should be installed according to the manufacturer's specifications and requirements in building or plumbing codes and local bylaws. Therefore, recommendations are based on general best practices and should be viewed more as minimum recommended requirements.

Do your best to ensure that every RHS component is easily accessible for inspection and maintenance.

Excavation

Before breaking ground, necessary permits and checks for buried lines or utilities have been taken out.

Requirements for excavation, bedding/support, backfilling, and ground cover have been checked with local codes and regulations and the product manufacturer's specifications.

Electrical

Installation of electrical connections, modifications to electrical circuits, and testing of components supplied with electricity are undertaken by a qualified professional or according to the manufacturer's instructions.

Permits

Any required permits have been obtained.

If you are installing an above-ground non-potable outdoor system, it is unlikely that you will need a permit.

Collection Area

The roof area is clear of debris.

Overhanging branches or sources of debris have been removed.

Pre-filter

If installing multiple pre-filters, they should be placed in order of coarsest to finest. Remove the largest debris first (e.g., leaf filters), then place your finest screen or filter last before the Transfer System.

Transfer System

The area between Collection Area and Storage Tank is clear of obstructions.

STEP 4: INSTALLATION AND TESTING

Pipes, fasteners/cement, and supports are the correct size and shape.

Any below-ground pipes have been properly excavated, supported, and backfilled according to the manufacturer's specifications and marked with tracer tape.

All sections of the Transfer System are structurally sound and do not have any unnecessary holes or other points of entry other than those required for water flow.

Transitions between components are sealed and watertight.

Storage Tank

Sufficient space for the Storage Tank to be delivered to the installation location.

Install location is level, stable, secure, and capable of supporting filled Storage Tank weight without shifting.

Install location is not prone to flooding or ponding.

Installation following manufacturer's instructions.

Direct sunlight is not shining into the Storage Tank.

Complies with local building codes and regulations.

There is clear access to a hatch or entryway for inspection and maintenance.

The Overflow pipe is directed away from the Storage Tank, buildings, and areas susceptible to erosion.

All inlets, openings, and outlets are sealed and watertight.

All hatches, vents, inlets, and outlets are either completely sealed or have screens.

Cold weather modifications are present if necessary.

Aboveground tanks are secure against wind and extreme weather even when empty.

Connections to buried tanks, including Transfer System, Overflow, Make-up, and any electrical conduits, have all been tested to ensure they are operational, properly sealed, and watertight before backfilling.

If you are concerned about aesthetics, ensure there is room around the Storage Tank for fencing, plants, or other ways to alter or hide its appearance.

Overflow

The Overflow system is installed below the Transfer System inlet to the Storage Tank.

The Overflow outlet pipe is directed away from other structures and will discharge water in an area that is not prone to erosion.

STEP 4: INSTALLATION AND TESTING

Connection System

Any below-ground pipes have been properly excavated, supported, and backfilled according to the manufacturer's specifications.

The Connection System is sealed and tested before hooking up the Pump System.

Pump System

If installed within the Storage Tank, the pump intake is located off the tank bottom; the recommended intake height is 4 inches (100 mm).

Any sensors or controls associated with the Pump System are in place and tested before operating the pump.

Pump Troubleshooting Checklist

- All powered components are connected to the electrical supply.
- Storage Tank water levels are higher than the outlet pipe to the Pump System.
- There are no blockages, faulty switches, or sensor issues.
- An automated bypass system has not been triggered.
- If pressure is too low, check that the pressure tank's static pressure is 2 psi (14 kPa) less than the desired cut-in (minimum) pressure.

- No frozen pipes.
- All shut-off valves are open.

Treatment System

Treatment System installed according to manufacturer's specifications.

Drinking water supplied by the treatment system has been tested for safety before use.

Fixtures

Fixtures are rated to use the flow and pressure of water supplied by the Pump System.

Excavating a backyard to install a RHS

TESTING CHECKLIST

Once installed, you must test your RHS to ensure it works as intended. This testing will be a more involved version of the

STEP 4: INSTALLATION AND TESTING

regular maintenance you will be performing later.

Even if you did not install the system yourself, ensure that your RHS has been appropriately tested before putting it into use, and request a demonstration from any qualified professionals who worked on your RHS.

Seals are Watertight

Around Pre-filters
Collection Area to Transfer System
Transfer System to Storage Tank
Storage Tank to Overflow
Storage Tank to Make-up
Storage Tank to Connection System
Connection System to Pump System
Pump System to Fixtures
Around all valves, pipe fittings, and openings
Around any cold weather diversions

Electrical Connections are Secure and Watertight

Pump System
Water Level Sensors
Control systems
Automated Make-up System
Automated Overflow System

Test Run

Water added to Collection Area makes it to Storage Tank at the expected collection efficiency.

The water level in the Storage Tank is stable.

Fixtures are receiving water at expected pressure and flow rates.

Water obtained from fixtures meets your desired quality.

STEP 5: MAINTENANCE

Now that your RHS is installed and operating, keeping a regular maintenance schedule will ensure it stays that way for years or decades. Use the following checklists to design a maintenance schedule for your RHS.

MAINTENANCE CHECKLIST

Since every RHS is different, these maintenance checklists should be used as a starting point for creating your own maintenance schedule.

When purchasing components, record all maintenance requirements and schedules recommended by the manufacturer in one place and use that information to adjust the checklist below to fit your system, climate, and intended use of rainwater.

Spring/Annual

- Divert Transfer System away from Storage Tank before cleaning or introducing water to your RHS
- Inspect and clean roof
- Prune overhanging branches
- Clean gutters and downspouts
- Clear any Collection Area Pre-filters
- Clear any Transfer System Area Pre-filters
- Clear any Storage Tank Pre-filters
- Check screens on Make-up and Overflow systems
- Check Storage Tank for any signs of water above the high water level
- Flush Transfer System
- Inspect Pump, reconnect if shut off for winter
- Check Pump System when no water is being used; if the pump cycles repeatedly, this may indicate that there are problems with the foot valve or check valve or that there is a leak
- Flush Connection System
- Inspect all fittings, seals, valves, and electrical connections
- Check for moist areas indicating leaks around all connections, joints, and openings
- Ensure that no float switches are tangled with other switches or wires
- Inspect sensors, controls, and any automated systems

STEP 5: MAINTENANCE

- Remove any dirt or debris on internal Storage Tank components
- Check sediment level in Storage Tank
- Remove sediment if more than 1/2 inch (13 mm) deep
- Check the security of the Storage Tank lock
- Change bulbs in UV Treatment System
- Inspect and clean filters in Treatment System
- Check warning lights or indicators
- Test water
- Reconnect Transfer System to Storage Tank

Summer

- Divert Transfer System away from Storage Tank before cleaning or introducing water to your RHS
- Clean gutters and downspouts
- Clear any Collection Area Pre-filters
- Clear any Transfer System Area Pre-filters
- Clear any Storage Tank Pre-filters
- Check warning lights or indicators
- Test water (if used as drinking water)
- Reconnect Transfer System to Storage Tank

Fall

- Divert Transfer System away from Storage Tank before cleaning or introducing water to your RHS
- Inspect roof surface
- Clean gutters and downspouts
- Clear any Collection Area Pre-filters
- Clear any Transfer System Area Pre-filters
- Clear any Storage Tank Pre-filters
- Inspect cold weather modifications
- Check warning lights or indicators
- Test water (if used as drinking water)
- Reconnect Transfer System to Storage Tank

Winter (Operational)

- Divert Transfer System away from Storage Tank before cleaning or introducing water to your RHS
- Clean gutters and downspouts
- Clear any Collection Area Pre-filters
- Clear any Transfer System Area Pre-filters
- Clear any Storage Tank Pre-filters
- Inspect cold weather modifications and check for ice
- Check warning lights or indicators
- Test water (if used as drinking water)
- Reconnect Transfer System to Storage Tank if operating through winter

STEP 5: MAINTENANCE

Decommissioning

- Drain all of the rainwater in the Storage Tank, Transfer System, and Connection System
- Shut off the water supply to the Make-up system
- Disconnect electrical supply to pumps, controls, and sensors
- Disconnect downspouts from the Transfer System and have them discharged to a suitable location
- Disconnect any Fixtures from the RHS and connect them to an alternate source
- Ensure all openings, inlets, and outlets are sealed to prevent contamination, insects, or animals from entering the RHS

Startup

- Ensure all openings, inlets, and outlets remained sealed while decommissioned
- Clean out disconnected pipes
- Disconnect any alternate sources and reconnect Fixtures to RHS
- Reconnect downspouts to the Transfer System
- Reconnect electrical supply to pumps, controls, and sensors
- Do a test run with a small volume of water
- Reconnect the Make-up system (if present)

Every Three Years

- Completely drain the Storage Tank, and inspect the entire RHS for damage or wear
- Clean out all sludge, sediment, or debris throughout the RHS

Cleaning out rain gutters

CLEANING AGENTS

Caution should be used when selecting cleaning agents. Harsh chemicals can damage your RHS or may introduce chemical contamination that must be thoroughly removed before rainwater can safely be captured.

A solution of hydrogen peroxide mixed according to the manufacturer's instructions is sufficient for heavy annual cleaning. While bleach or other chlorine-based cleaners are often recommended, these come with their own risks, and it is not necessary to expose yourself to these dangerous chemicals.

Scouring agents, heavily acidic or basic chemicals, or other cleaning agents are not recommended unless specifically identified by the manufacturer of that component as being safe to use.

> *Important*: It can be very difficult to flush out and completely remove any chemicals you introduce into your RHS, so use caution before introducing anything new.

ADDITIONAL EVENTS

If your area has seasonal sources of contaminants that can end up on your roof, you may want to schedule some additional maintenance events to deal with these. Examples could be:

- Pollen season in spring
- Trees shedding leaves or needles in fall
- Increased bird activity during migration or nesting periods
- Dust from wildfire smoke
- Nearby construction or roadwork
- Seasonal increases in insect or pest activity

Storage Tank Cleaning

In some cases, removing the accumulated sediment at the bottom of your Storage Tank may be necessary. Pumps capable of handling large debris and solids (for example, sump pumps, effluent pumps, or wet vacuums) can be used to pump out this sediment layer. However, regularly removing sediment or cleaning the Storage Tank is not recommended because this can destroy beneficial 'biofilms' in the tank. These biofilms may contribute to improved stored rainwater quality.

COMMON PROBLEMS AND TROUBLESHOOTING

Bad smelling water

Smells are usually caused by algae growth, sediment buildup, or dead animals in the Storage Tank.

Cleaning out and disinfecting the Storage Tank will solve this problem. To prevent reoccurrence, ensure that your Pre-filter systems are installed and working correctly to reduce sediment buildup, and that there are no holes or points of entry for animals in your RHS.

Harvesting less water than expected

Blockages or leaks usually cause this situation. You can trace leaks by manually adding water to the Collection Area, but blockages can be more challenging to find. If you can't find any obvious culprits (leaves, stones, or sediment) in common blockage locations (transitions between RHS components, bends, or outlets), you may have to access cleanouts and manually add water at specific points to see where flow is being disrupted. If parts of your RHS are regularly blocked, you should alter the design or add Pre-filters at strategic locations.

Make sure to double-check your calculations for expected water capture. Check against actual values for recent precipitation and ensure you have factored in various inefficiencies like first-flush systems and the proportion of your roof used for the Collection Area.

Muddy or unclear water

If your harvested water looks brown or grayish, sediment or debris is getting into your water. Add more filters with finer screens if possible, and check that your filters are working and that sediment has not accumulated in your Storage Tank. However, filters can accumulate dirt and debris over time, and a dirty filter will eventually pass through low-quality water. A potential upgrade to resolve this could be the addition of a downpipe filter. This additional filter is placed within the Transfer System to remove debris before reaching the Storage Tank. Think of it like an advanced Pre-filter.

One possible solution is to pass the rainwater through a sand filter system before reaching the Storage Tank. This filter can significantly improve your water quality but can be complicated and may require a qualified professional for proper design and installation.

Another cause could be accumulated sediment or sludge at the bottom of your tank. During heavy rainfall, water coming in from the Transfer System could disturb this sediment layer and result in poor quality water. One option is to use a calmed or U-shaped pipe to end your Transfer System within the Storage Tank. The Transfer System pipe enters the Storage Tank, extends near the bottom, and ends in an upwards-facing u-shape.

This modification allows rainwater to enter your Storage Tank at a slower speed so that it doesn't create turbulence and stir up the sludge at the bottom.

Rainwater for the garden

EXAMPLE SYSTEMS

The basic outdoor rainwater harvesting system shown as an example in this chapter is a simple plumbing job that could be a DIY weekend project.

The large outdoor system example is more involved, and even though it may require some additional hired help, it is something you can now completely design yourself with the tools available in this book.

For any more complex system, or use of rainwater for indoor applications, you should have your design reviewed by qualified professionals to ensure it meets all local codes and requirements and everything is installed safely and securely.

BASIC OUTDOOR SYSTEM

Watering lawns and gardens are often the single biggest use of a household's water. So even a basic outdoor system can cut down your water bill substantially. This RHS is designed to use as few additional components as necessary, and the total cost will be a few hundred dollars and an afternoon or two of your time.

The Transfer System will use your existing gutters and downspouts to move rainwater to a barrel-type aboveground Storage Tank, which can be used for non-potable outdoor uses without additional treatment.

Creating a solid surface for your Storage Tank to sit on and cutting or replacing the existing downspout device will be all the construction necessary. A passive Overflow pipe will divert excess water from your home if the barrel gets too full.

Basic RHS installed and working

Materials

- 55-gallon plastic barrel
- Inline downspout diverter
- Spigot (if not included with barrel)
- Plumber's tape
- A few bags of sand
- 2 concrete pavers or stones to cover a 2 sq. ft. area
- A small section of pipe or hose to direct overflow

Tools

- Drill
- Adjustable wrench
- Hacksaw
- Hole saw, 1-1/4-inch diameter
- Utility knife
- Shovel
- Level
- Marker

Select a space

You will be placing a rain barrel directly under an existing downspout. If you have multiple downspouts, see which one has the largest water flow during the next rain shower. You will also want to be near plants or lawn spaces that need regular watering.

Preparation

Once you have a spot picked out, you will need to ensure it is stable and can support the weight of a full barrel. A hard surface like a patio or concrete slab will work great.

If you want to put your rain barrel onto a surface that is currently grass-covered, you will need to make some modifications. Use a shovel to dig about 6 inches (15 cm) into the dirt, backfill with sand, then top with concrete slabs or stones. Use the level to make sure this surface is completely flat.

Find a good spot nearby for overflow. Since you are diverting a downspout, you should direct any overflow to wherever your downspout previously discharged. An Overflow only requires a short section of pipe or hose.

> *Optional:* Elevating the rain barrel using a wooden platform or any other stable structure can make it easier to use. Since your spigot will be on the bottom of the barrel, raising it a foot or two might make your life easier in the long run.

Purchase your RHS

Select a rain barrel that meets the following requirements:

- It has a removable cover that can be tightly sealed

- It has an inlet to attach to an inline diverter near or at the top
- Has an attached spigot or garden hose connection
- It has an outlet near the top for an overflow pipe
- Will fit securely in a space directly beside or below your downspout
- Is made of food-grade material or meets NSF/ANSI 61 Standards (*optional*)
- Looks great in your yard (*optional*)

Select an inline diverter that meets the following requirements:

- Can attach to your existing downspout (materials and size are compatible)
- Can connect to your selected rain barrels inlet

Modify the downspout

- Mark a location on the downspout at least 24 inches (60cm) higher than the top of the rain barrel (account for any additional height from a platform the barrel might be sitting on)
- Cut the downspout and attach the inline diverter to the top and bottom portions; follow any manufacturer's instructions that come with the diverter.

Modify the barrel

- If your barrel does not have a built-in spigot, you will need to drill a hole to attach one.
- Drill a hole at least 2 inches (50 mm) from the bottom of the barrel, and clean up the edges with a knife. The hole size will be based on the size of the spigot.
- Attach the spigot and gaskets, and secure them with plumbers' tape. Make sure the spigot is pointing down.
- If your barrel has no built-in overflow, you will need to drill a hole to attach one.
- Drill a hole 4 inches (100 mm) from the top of the barrel, and clean up the edges with a knife. The size of the hole will be based on the diameter of the outflow pipe or hose you selected. Remember, the Overflow should be the same size or larger than the Transfer System.
- Attach the overflow pipe and direct it to a suitable discharge location, likely near the existing downspout opening

Attach diverter to barrel

- Connect the hose or pipe from the inline diverter to the inlet near the top of the rain barrel.

- Test the system by spraying water onto your roof and ensuring all your connections are secure and rainwater makes it into your RHS.

Additions to the Basic System

These are some easy changes that can improve the quality and useability of the basic RHS. Most of these additions can be made after the initial installation and take minimal cost or effort to include:

- Storage Tank expansion: if your barrel is regularly overflowing or if you need to capture even more water, you can connect two (or more) barrels. Just add a connecting pipe or host below the existing overflow. This option makes the basic system the best choice for most homeowners, as expanding to 2, 3, or more barrels can significantly increase capacity without significant renovations.
- Pump: adding a small hand-powered or electric pump to your system can increase the water pressure and make it much easier to use your harvested rainwater. These can either fit inside the barrel or sit discretely to the side.
- Pre-filter: adding gutter guards or leaf filters to the gutters which feed the downspout connected to your RHS will prevent clogs and make maintenance and cleaning of your RHS much easier.

- Easy overflow: connect your overflow pipe directly back into the downspout to simplify your RHS and give it a cleaner look.

LARGE OUTDOOR SYSTEM

Creating an extensive non-potable outdoor system can be more challenging; however, even if you have to hire professional help, you will be in a better place to direct these professionals if you understand your goals and realistic ways to achieve them. As the size of an RHS increases, the following issues come up.

Basic systems may only require a few hours of quick planning and design, while larger systems might require multiple design iterations and more serious thought. Realizing that you can't achieve your initial goals based on hard limits or design constraints and going back to the drawing board if you can't source the right parts within your budget is a normal part of the process if you are new to rainwater harvesting.

Any complex construction project will involve more time spent designing, compromising, and resetting expectations than assembling the pieces.

EXAMPLE SYSTEMS

Large RHS with multiple Storage Tanks

For the most common issues that come up as you design a larger RHS, a few potential technologies or solutions are recommended below:

Problem: Insufficient Collection Area to meet demand

Solution: If only a portion of the roof surface is used as a Collection Area, you can expand the Transfer System to incorporate the entire roof surface. This expansion may require additional piping and connections in your Transfer System.

Solution: A secondary structure, such as a nearby shed, garage, or new structure dedicated to the RHS, can be incorporated into the Transfer System. Multiple Collection Areas can feed into the same Storage Tank, but your Overflow will need to be able to accommodate the total potential inflow from all sources.

Problem: Lack of space for larger Storage Tank

Solution: Buried Storage Tanks are a standard solution for lack of space. It is unnecessary to bury the entire Storage Tank, and partially buried tanks can also save space. There are additional options for Storage Tank shapes, such as thin tanks (that fit up against exterior walls), roof-mounted tanks (which will need sign-off from an engineer), or tanks that can also be placed indoors in garages or basements.

Solution: Similar to the Basic Outdoor System described earlier, smaller Storage Tanks can be linked or chained together into a more convenient shape than a single large tank.

To keep this solution simple, ensure all Storage Tanks are placed at the same elevation.

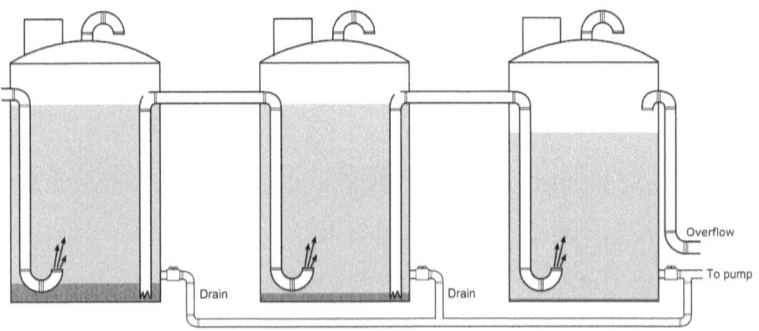

Multiple Storage Tanks chained together

Problem: Fixtures are far away or require a Pump System for pressure

Solution: The most straightforward method is an all-in-one pump solution that attaches directly to or sits inside your RHS. Pump system packages designed for an RHS may be slightly more expensive but will have compatible parts, and you can install them yourself.

Solution: If getting an electrical connection is a problem, incorporating a hand pump solution is a low-cost and relatively low-maintenance option. However, you must manually pump into the pressure tank whenever you need water.

Solution: Elevating your Storage Tank can provide a small boost of additional pressure from gravity; however, every foot of elevation only gives you about 0.4 PSI. Realistically, you need about 30 feet (10 m) of height to get decent water pressure this way.

Problem: Water quality is poor, and a Treatment System is needed

Solution: A Pre-filter will increase the effectiveness of any Treatment System, and installing one should be the first step in improving water quality. The most accessible options are gutter guards and leaf screens; however, the best option is to incorporate a first-flush diversion into the Transfer System.

Solution: All-in-one Treatment System packages intended to connect directly to the RHS or set up at the fixture needing treated water are the most straightforward approach. Filtration systems can be relatively straightforward to install and maintain. Designing a chemical treatment system and determining the correct chlorine or disinfectant application rates can be difficult and time-consuming, and is not recommended for non-potable use.

Solution: Adding a slow-sand filtration system between your Transfer System and Storage Tank will ensure that water entering your tank is as high quality as possible. The downsides of this option are that these sand filtration systems may require a lot of space and have additional maintenance needs. However, the upside is that your rainwater may now be such high quality that it meets drinking water regulations and could be potable.

CONCLUSION

Congratulation! You now have all the information you need to design, install, and maintain a rainwater harvesting system for your home.

But don't stop now, get started and create your version of the Basic system right away, or commit to designing a more complex RHS that will change your relationship with water.

This is a small change that benefits you and your family, which also benefits the environment. Actions like this improve more than just our own lives, they set an example for others and help steer our communities in the right direction.

If you enjoyed *Harvesting Rainwater for your Home* please don't be shy about saying so! My goal is to help as many people as possible appreciate the environment around us, so feel free to share this book with your friends and family (and please leave a review wherever you purchased this book).

GLOSSARY

Aboveground Storage Tank - any tank, barrel, or cistern that is not partially or fully buried

Air Gap - a physical separation between two systems to prevent backflow, usually represented by a gap between pipes

Backflow - water flowing the wrong direction in a pipe, for an RHS generally refers to water returning from the Connection System to the Storage Tank or from the Storage Tank back into a Make-up System

Back-up System - see Make-up System

Backwater Valve - a one-way valve to prevent water from flowing in the wrong direction in a pipe

Catchment - is an area that catches or collects water, in this case rainwater or other precipitation

Cistern - see Storage Tank; cisterns are typically large and usually installed below ground

Cleanout - an easily accessible section of pipe with a removable cover that allows for inspections, maintenance, or cleaning

Collection Area - any surface used to collect falling rain and which is connected to a Transfer System

Confined Space - any space with limited or restricted means for entry or exit that is not designed for continuous occupancy; large Storage Tanks can be classified as Confined Spaces

Conveyance Network - see Transfer System

Cross-connection - any actual or potential connection between a potable water system and any source of pollution or contamination, like an RHS

Dead Space - the volume of water below the low water mark in a Storage Tank, below which a pump can reach, so this water is inaccessible

Domestic Use - Any indoor use of water that could intentionally or accidentally result in direct contact (e.g., showering, bathing) or ingestion (e.g., drinking, cooking) of water; this is considered equivalent to drinking water for regulatory purposes by most jurisdictions

Downspout - vertical pipes connected to gutters that carry rainwater to the ground and away from the building

Drinking water - any water of sufficient quality or purity so that people can safely consume it

Endcap - a cap that pushes over the outside or threads onto the end of pipes to create a watertight seal

Filtration - the process of physically separating contaminants from water; filters can be meshes, screens, or membranes

Fixture - any part (e.g., sink, toilet, shower) or appliance (e.g., dishwasher, laundry) that uses water or is attached directly to a water supply system

Float switch - can detect the level of a liquid in a tank or container by floating and acts as a mechanical switch triggered as the water level goes up or down

Freezing depth - see Frost penetration depth

Frost line - see Frost penetration depth

Frost penetration depth - the maximum depth to which frost or frozen groundwater is expected to reach, determined by a local regulatory or governing authority

Gray water - is used water from baths, showers, or sinks but doesn't include toilet, kitchen, or dishwasher waste; it may be used for low-risk purposes such as below-ground irrigation or toilet flushing

Gutter - an open pipe along the edge of a roof that carries rainwater away from a building

Gutter guard - sits on top of the gutter, allowing water to pass through but not leaves or other objects; the size of the holes in the screen, filter, or mesh on the gutter guard can vary

Hanger - attaches a gutter to a roof

HDPE - high-density polyethylene, a type of heavy-duty plastic

Head - is a measurement of pressure and is the maximum height that a pump can move water vertically against gravity

High water level - the highest point to which your Storage Tank can be filled; completely filling your tank can damage your RHS, so an Overflow pipe will remove any water above this point

In-line diverter - attaches to the downspout and redirects a portion of the water into the Storage Tank

Leaching - is when chemicals or other contaminants release from solid surfaces into water

Leaf Guards - See gutter guards

Low water level - the minimum water level in your Storage Tank which allows the Pump System to operate

Make-up System - a manual or automatic system to add water to the Storage Tank, preventing the water level from dropping below the low water level

Manifold - an indoor plumbing distribution system

Non-potable - any use of water that is not intended for human consumption

NSF/ANSI - National Sanitation Foundation standards developed by the American National Standards Institute, an internationally recognized standard development organization

NTU - nephelometric turbidity units, a measure of turbidity

Overflow - collected rainwater in excess of Storage Tank capacity

Parapet Gutter - a gutter with a short wall behind it, located between the roof surface and a parapet

Potable - water intended to be used for human consumption (drinking or cooking)

Pre-filter - any addition to the Collection Area or Transfer System intended to prevent contamination of the harvested rainwater, typically is a form of filter, screen, or guard

Soakaway Pits - excavated pits will with gravel or similar material that allow excess rainwater to drain back into the soil slowly

PVC - polyvinyl chloride, a type of plastic pipe

Storage Capacity - the volume of a Storage Tank between the low water mark and the high water mark

TDS - Total Dissolved Solids, includes everything dissolved in the rainwater and is not visible; this can consist of chemicals and small dust particles.

Top-up System - see Make-up System

Transfer System - a series of pipes moving harvested rainwater from the Collection Area to the Storage Tank

Treatment - any individual or combination of components that improve the quality of harvested rainwater, typically installed after the pump, but passive filters and settling tanks may be placed before the Storage Tank

TSS - Total Suspended Solids, includes everything too large to be dissolved into the rainwater, bits of dirt, dust, and chemicals that don't dissolve in water; as a result, they make your water cloudy.

Turbidity - is a measure of how clear the rainwater is; higher turbidity is a sign that your rainwater contains more suspended solids

Underground Storage Tank - a Storage Tank that has been buried

UPVC - is unplasticized polyvinyl chloride, a strong, low-maintenance, and lightweight plastic

UNIT CONVERSION CHEAT SHEET

Distances

1 inch = 25.4 mm = 2.54 cm

1 cm = 10 mm = 0.394 inches

1 foot = 0.305 m

Areas

1 square foot (sq. ft) = 0.093 square meters (sq. m)

1 square meter = 10.8 square feet (sq. ft)

Volumes

1 US gallon (USG) = 0.83 Imperial Gallons = 3.79 liters

1 liter = 0.26 US gallons

1 cubic foot = 7.48 US gallons = 0.028 cubic meters

1 cubic meter = 1000 liters = 35.3 cubic feet

Weights

1 cubic foot = 62.5 pounds

1 USG = 8.35 pounds

1 cubic meter = 1,000 kg

ANNUAL RAINFALL BY STATE

Use this section to get a rough idea of annual precipitation by state. Check local weather information sources for a complete picture of when the wettest months are and what to expect for worst-case storm events, but this can work as a starting point.

State - Average yearly precipitation (inches)
Alabama - 58.3
Alaska - 22.5
Arizona - 13.6
Arkansas - 50.6
California - 22.2
Colorado - 15.9
Connecticut - 50.3
Delaware - 45.7
Florida - 54.5
Georgia - 50.7
Hawaii - 63.7
Idaho - 18.9
Illinois - 39.2
Indiana - 41.7
Iowa - 34
Kansas - 28.9
Kentucky - 48.9
Louisiana - 60.1
Maine - 42.2
Maryland - 44.5
Massachusetts - 47.7
Michigan - 32.8
Minnesota - 27.3
Mississippi - 59

ANNUAL RAINFALL BY STATE

Missouri - 42.2
Montana - 15.3
Nebraska - 23.6
Nevada - 9.5
New Hampshire - 43.4
New Jersey - 47.1
New Mexico - 14.6
New York - 50.3
North Carolina - 17.8
North Dakota - 17.8
Ohio - 39.1
Oklahoma - 36.5
Oregon - 27.4
Pennsylvania - 42.9
Rhode Island - 47.9
South Carolina - 49.8
South Dakota - 20.1
Tennessee - 54.2
Texas - 28.9
Utah - 12.2
Vermont - 42.7
Virginia - 44.3
Washington - 38.4
West Virginia - 45.2
Wisconsin - 32.6
Wyoming - 12.9

ADDITIONAL SOURCES

American Rainwater Collection Systems Association (ARCSA). 2009. Rainwater Catchment Design and Installation Standards. *Revised Dec. 15, 2010.* www.arcsa-usa.org.

Australia Rainwater Industry Development Group (ARID). 2009. Rainwater 2009 Consumer Guide & 2008 Consumer Guide. www.arid.asn.au

Building Capacity for Rainwater Harvesting in Ontario: Rainwater Quality and Performance of RWH Systems. 2008. Despins, C. M.Sc. Thesis, University of Guelph.

Goyal, R. 2014. Rooftop Rainwater Harvesting: Issues and Challenges. https://www.researchgate.net/publication/283150765_Rooftop_Rainwater_Harvesting_Issues_and_Challenges.

Regional District of Nanaimo. 2022. Rainwater Harvesting Best Practices Guidebook.

Texas Water Development Board. 2005. The Texas Manual on Rainwater Harvesting. http://www.twdb.state.tx.us/home/404.asp.

IMAGE CREDITS

Introduction: Garden rain barrel (image by schulzie under the CanvaPro License)

Step 1: Determining Collection Area size (image from https://rainwaterharvesting.tamu.edu/)

Step 2: Overflowing Gutter (image by cogdogblob under Creative Commons Attribution 2.0 License)

Step 2: Pre-filter before the downspout (image from http://agrilife.org/rainwaterharvesting)

Step 2: Simple first-flush system (image from http://agrilife.org/rainwaterharvesting)

Step 2: Example underground dry system (image from Rainwater Harvesting Best Practices Guidebook, Regional District of Nanaimo)

Step 2: Frozen downspout pipes (image by Natalia Nosove under CanvaPro License)

Step 2: Concrete below ground storage taken (image by olejx under CanvaPro License)

Step 2: Storage Tank on elevated platform (image by SuSanA Secretariat under the Creative Commons Attribution 2.0 License)

Step 2: Water pump (image by supersmario under CanvaPro License)

Step 2: Simple rainwater overflow (image by tanyss under CanvaPro License)

Step 2: UV disinfection system (image from https://rainwater-harvesting.tamu.edu/)

Step 2: Example potable water treatment system (image from http://agrilife.org/rainwaterharvesting)

Step 2: Simple in-home water treatment system (image by nataistock under CanvaPro License)

Step 3: Common styles of material certification logos (image from http://agrilife.org/rainwaterharvesting)

Step 3: Determining operational Storage Tank capacity (image from Rainwater Harvesting Best Practices Guidebook, Regional District of Nanaimo)

Step 3: Simple pump in a small Storage Tank (image from Rainwater Harvesting Best Practices Guidebook, Regional District of Nanaimo)

Step 4: Excavating a backyard (image by valentymsemenov under the CanvaPro License)

Step 5: Cleaning the gutters (image by Fokusiert under CanvaPro License)

Step 5: Rainwater for watering the garden (image by Astrid860 under CanvaPro License)

Example System: Basic rain barrel system (image from www.pwdraincheck.org)

Example System: Large rainwater tanks (image by zstock-photos under CanvaPro License)

Example System: Multiple Storage Tanks chained together (image from http://agrilife.org/rainwaterharvesting)

www.ingramcontent.com/pod-product-compliance
Lightning Source LLC
Chambersburg PA
CBHW031112080526
44587CB00011B/944